Introduction to Global Energy Issues

Introduction to Global Energy Issues

Renaud Gicquel
Mines ParisTech, Paris, France

May Gicquel
Ministry of Economy and Finance, Paris, France

CRC Press
Taylor & Francis Group
Boca Raton London New York

CRC Press is an imprint of the
Taylor & Francis Group, an **informa** business
A BALKEMA BOOK

Originally published in French.

First edition:
'Introduction aux Problèmes Energétiques Globaux' by Renaud Gicquel
© 1992, Economica, Paris

Second edition:
'Introduction aux Problèmes Energétiques Globaux' by Renaud Gicquel & May Gicquel
© 2013, Presses des Mines, Paris

Published by:
CRC Press/Balkema
P.O. Box 447, 2300 AK Leiden, The Netherlands
e-mail: Pub.NL@taylorandfrancis.com
www.crcpress.com – www.taylorandfrancis.com

ISBN 13: 978-1-138-00014-8 (hbk)

Typeset by V Publishing Solutions Pvt Ltd., Chennai, India

Library of Congress Cataloging-in-Publication Data

Gicquel, Renaud.

[Introduction aux problèmes Energétiques globaux. English]
Introduction to global energy issues/Renaud Gicquel, Mines ParisTech, Paris, France, May Gicquel, Ministry of Economy and Finance, Paris, France
pages cm
Translation of: Introduction aux problèmes Energétiques globaux/Renaud Gicquel. 1992.

Summary: "This book seeks to provide a basis for reflection on energy at a global level. It synthesizes the main energy-related issues and requires no specialist knowledge, either of economics, engineering sciences or international relations. It comprises four main chapters, that discuss the major energy sectors (oil, natural gas, coal, synthetic hydrocarbons, electricity, renewable energy), the impacts of energy at macro-economic level, geopolitics and new issues related to energy, notably environmental aspects and successes of demand management policies. The book is aimed at industry professionals, (under) graduate students/academics, economists and policy-makers in the energy sector/field"–Provided by publisher.

Includes bibliographical references and index.
ISBN 978-1-138-00014-8 (hardback)
1. Power resources. 2. Energy policy. I. Gicquel, May. II. Title.

TJ163.2.G5313 2013
333.79–dc23 2013018461

To Fawzia, without whom this book would not exist

Contents

Preface

Jean-Arnold Vinois, Honorary Director at the European Commission, Former Director of the internal energy market within the Directorate General for Energy.

The book you have in front of you is an update of the first edition published in 1992, i.e. a little more than twenty years ago. If the 1990s were not exceptionally significant in terms of energy developments, we cannot say the same of the third millennium, which has seen energy issues move to the very forefront of current affairs and become a major component of household budgets and business competitiveness. This meant that it was time to refresh both the contents of this textbook for students as well as the interactive teaching methods that are proposed by the authors in the introduction and in Annex 1. And it is truly pedagogy that we are dealing with here. Each chapter of this book alone could be transformed into a book numbering 300 pages and it is no small achievement of the authors that they have succeeded in synthesizing many concepts, principles and additional information within a very limited space, while keeping things simple and accessible to any reader. Specifically, beneath the simple exterior lies a highly complex subject matter that is evolving ever more quickly under the influence of new technologies and new policies dictated by environmental imperatives and far-reaching geopolitical changes.

Fukushima, Macondo, Sandy are infamous names directly related to the energy problem, which alarmed the whole world and reminded man that energy can be unpredictable and indomitable.

The Fukushima event, in March 2011, highlighted our vulnerability to nuclear disasters and showed that the damage can be so excessive that no one country can bear the burden. This event has dramatically challenged the energy mix of countries like Germany, Switzerland, Belgium and even France.

Macondo reminded us that deep sea oil exploration presents major risks that must be contained by all stakeholders in the sector. Environmental damage caused by the accident which took place in April 2010 in the Gulf of Mexico has been so extensive that putting it right almost resulted in the demise of BP. Never before has compensation for an industrial accident of this type peaked so high. The dominance of oil in today's economy, its role in producing countries, the cost to consumers, risks of shortages and geopolitical importance are described in detail in this book, providing critical data on what will probably remain for a number of decades our main source of energy.

Sandy, which ravaged the east coast of the United States in November 2012 and especially New York and its financial center, reminded us again that climate change is at work. For the scientific community, it is the result of human activities, particularly greenhouse gas emissions produced by these activities, be they transportation, power generation and other industrial activities or simply human lifeways. The combination of energy and climate is undoubtedly the main cause of the temperature increase on Earth causing disturbances that we are experiencing ever more frequently. This is why the European Union has adopted a proactive policy to reduce greenhouse gas emissions between 2007 and 2020 (by 20% as compared to 1990), to increase use of renewable energy sources (up to 20% of the energy mix) and to improve energy efficiency (by 20% in 2020 as compared to the "business as usual" scenario).

The reader of this book will once again find all the basic information that will enable him/her to understand the forces at play, whether demographic change, economic growth, the use of certain forms of energy, environmental impacts, and even including the fuel wood crisis in developing countries (chapter 6.4.5), while being informed of the policies (in the process of being) implemented.

The authors should be applauded for bringing within reach of everyone all the essential concepts that govern the energy sector and for placing them in a broader perspective because energy lies at the heart of economic development and political governance. It is probably true that the planet has enough resources to meet growing demand but the authors rightly introduce all the factors that explain why energy policy will always be difficult to decide upon. The energy mix is probably the most difficult to harmonize with the environmental dimension, in particular because of the long time required to change the composition of the mix. The European Union, as proactive as it may be, has only reduced its dependence on fossil fuels from 81% to 75% between 1998 and 2011. At this rate, how many more decades will be required to eliminate greenhouse gas emissions entirely?

There is no doubt that the environmental dimension will play an increasing role in the evolution of energy use. Just ask the people of Beijing and Shanghai what they thought of January 2013, when these cities were dominated by the "smog", a combination of fog and smoke pollution. Economic growth can only be achieved by taking full account of this dimension and the internalization of external costs such as environmental costs will eventually prevail. Behavior will also change and demand management will soon be revolutionized by the implementation of information technology throughout the chain of production, transport and distribution of electricity. Avoiding energy consumption is now the first priority.

The European Union bases its fast-evolving energy policy (see Section 7.12) on three pillars: competitiveness and affordability, sustainability and supply security. The balance between these three pillars must constantly be maintained in an environment subject to many variables. Understanding what is happening, why and how, is the first step in the definition of policies. It is to the credit of the authors of this book that they have contributed with modesty and competence to this need. No doubt students, but also any other readers, can confidently draw on this "little" book to feed their debates on the fundamental issues of our time in the field of energy. This book's ambition has undoubtedly been to whet our appetite, and I think I can say that it has succeeded in doing so.

Acronyms

3D	Three-Dimensional
AACAES	Advanced Adiabatic CAES
API	American Petroleum Institute
ASPO	Association for the Study of Peak Oil and Gas
BANANA	Build Absolutely Nothing Anywhere Near Anything
BWR	Boiling Water Reactors
CAES	Compressed Air Energy Storage
CAP	Chilled Ammonia Process
CC	Combined Cycle
CCS	Carbon Capture and Storage
CDM	Clean Development Mechanism
CFC	Chlorofluorocarbons
CIF	Cost, Insurance and Freight
CIS	Commonwealth of Independent States
CNG	Compressed Natural Gas
COP	Conferences of the Parties
CT	Combustion Turbine
DC	Developing Country
DOE	Department of Energy
EC	European Commission
ECSC	European Coal and Steel Community
EEC	European Economic Community
EGR	Exhaust Gas Recirculation
EOR	Enhanced Oil Recovery
EPFL	École Polytechnique Fédérale of Lausanne
EPR	European Pressurized Reactor
ETS	Emissions Trading Scheme
EU	European Union
EURATOM	European Atomic Energy Community
FAO	Food and Agriculture Organization of the United Nations
FOB	FOB
FPI	Food Processing Industries
GDP	Gross Domestic Product
GFR	Gas Fast Reactors
GIF	Generation IV International Forum

GIZ	Deutsche Gesellschaft für Internationale Zusammenarbeit
GNI	Gross National Income
GNP	Gross National Product
GWP	Gross World Product
GWP	Global Warming Potential
GT	Gas Turbine
GT-MHR	Gas Turbine Modular Helium-cooled Reactor
HCFC	Hydrochlorofluorocarbons
HDI	Human Development Index
HFC	Hydrofluorocarbons
HHV	Higher Heating Value
HPE	Haute Performance Energétique (high energy performance)
HPES	Hydro Pneumatic Energy Storage
HTR	High Temperature Reactors
IEA	International Energy Agency
IEPE	Institut d'Économie et de Politique de l'Énergie
IFP	French Petroleum Institute
IGOs	Intergovernmental Organizations
IPCC	Intergovernmental Panel on Climate Change
IPP	Independent Power Producers
ITER	International Thermonuclear Experimental Reactor
ITH	Integration Time Horizon
LCA	Life Cycle Analysis
LDCs	Least Developed Countries
LFR	Lead Fast Reactors
LHV	Lower Heating Value
LNG	Liquefied Natural Gas
LPG	Liquefied Petroleum Gas
MRV	Measuring, Reporting and Verification
MSR	Molten Salt Reactors
NCG	Non-Condensable Gas
NGL	Natural Gas Liquids
NGOs	Non-Governmental Organizations
NICs	Newly Industrialized Countries
NIMBY	Not In My Back Yard
NUGG	Natural Uranium Graphite Gas
ODA	Official Development Assistance
OECD	Organization for Economic Cooperation and Development
ODP	Ozone Depletion Potential
OPB	Oleo-Pneumatic Battery
OPEC	Organization of Petroleum Exporting Countries
OTEC	Ocean Thermal Energy Conversion
PBMR	Pebble Bed Modular Reactor
PPP	Purchasing Power Parity
PURPA	Public Utility Regulatory Policy Act
PV	Photovoltaic
PWR	Pressurized Water Reactors

R & D	Research and Development
R/P	Ratio of Proven Reserves to Annual Production
RE	Renewable Energy
RUE	Rational Use of Energy
SCWR	Supercritical Water Reactors
SFR	Sodium Fast Reactors
STL	Storage of Latent heat
SWU	Separation Work Unit
TEWI	Total Equivalent Warming Impact
TFEU	Treaty on the Functioning of the European Union
THESE	Thermal Energy Storage of Electricity
TIPP	French domestic tax on petroleum products
UNCTAD	United Nations Conference on Trade and Development
UNDP	United Nations Development Programme
UNEP	United Nations Environment Programme
UNFCCC	United Nations Framework Convention on Climate Change
UNIDO	United Nations Industrial Development
USA	United States of America
USSR	Union of Soviet Socialist Republics
VAT	Value Added Tax
VHV	Very High Voltage
VOCs	Volatile Organic Compounds
WEC	World Energy Conference
WHO	World Health Organization
WMO	World Meteorological Organization

About the authors

The main part of Renaud Gicquel's curriculum has taken place at École des Mines de Paris (Mines ParisTech), at its Centre for Energy Studies which he headed from September 1987 to January 2003. Named full professor in July 1986, he teaches applied thermodynamics, an introduction to global energy issues and energy system modeling.

He was formerly Special Assistant to the Secretary General of the United Nations Conference on New and Renewable Sources of Energy from April 1980 to October 1981 in New York and Chargé de mission for multilateral issues with the Service des Affaires Internationales of the Ministry for Research and Technology in Paris in 1982. From 1983 to 1985, he was Adviser for International Issues at the Centre National de la Recherche Scientifique (CNRS). In 1986, he was one of the two founders of the Observatoire Méditerranéen de l'Énergie (OME). From May 1991 to May 1994, he was Deputy Director of the École des Mines de Nantes (EMN) in charge of Research, and head of its Energy Systems and Environment Department.

He has published about 80 papers in the field of energy and two books:

- in 1992 "Introduction aux Problèmes Énergétiques Globaux" (Editor: Economica, Paris);
- in 2001 "Systèmes Énergétiques, Tome 1 et Tome 2" (Editor: Presses des Mines, Paris). A revised edition and a third volume were published in 2009.

Revised versions of these books have been translated into English and published by CRC Press:

- "Energy Systems, a new approach to engineering thermodynamics" in 2012;
- "Introduction to Global Energy Issues" in 2013.

Renaud Gicquel created the Thermoptim-UNIT portal (www.thermoptim.org) dedicated to a new paradigm for teaching applied thermodynamics, as well as various software packages, in particular Thermoptim, which is used by over one hundred and twenty higher education institutions worldwide.

May Gicquel graduated from École Polytechnique – ParisTech as an engineer in 2007 and subsequently joined the Corps des Ponts, des Eaux et des Forêts to become a high-ranking civil servant in France.

From 2008 to 2010, her work focused on emerging economies (mainly China and India). She first studied the impact of the arrival on the international scene of firms from these economies on the global automotive sector and on its industrial organization, in terms of innovation, partnerships, and business models. She also researched how these countries' extractive industries (coal, iron ore, aluminum, etc.) were restructuring to enter the global competition. The following year, she was based in Beijing to survey the development of electric and hybrid vehicles in China, taking into account the public policies that were being implemented, the new technologies under R & D, the growing number of industrial players, and the definition of Chinese standards and security tests.

In 2010, she returned to Paris to work for the French Ministry of Finance's Directorate-General of the Treasury (DG Trésor), where she was in charge of supervising the AFD group's strategy and finances for three years. AFD (the French Agency for Development) is the agency responsible for implementing most of the French bilateral official development assistance (ODA).

In 2013, she joined the Treasury's Eurozone Task Force, focusing on the European Stability Mechanism.

Chapter 1

Introduction

1.1 OVERALL PRESENTATION

From the discovery of fire to that of the atom, the development of human societies has largely been based on the conquest of energy. In all countries, energy has gradually become one of the key factors of social and economic development, along with capital, labor and natural resources. It is now impossible to do without it.

After decades of cheap energy flowing almost freely, crises have become the rule for the last forty years. This disruption of the energy landscape is of particular concern as the impact of energy crises on human societies can be considerable:

- economic recession;
- rising unemployment;
- international conflicts as a result of oil shocks;
- shortage of traditional fuels as a result of the firewood crisis;
- reduction of the ozone layer;
- increase of the greenhouse effect and acid rains as a result of energy emissions.

This book seeks to provide a basis for reflection on all global energy problems, offering an analysis of the main aspects to consider: energy supply, resource-dependent industries and available technology, macroeconomic implications of energy demand, geopolitical issues, and specifics of the situation in developing countries. It does not thoroughly address environmental issues, which would require a study beyond the limits we set.

This book is the second edition of a book published in 1992, at a time when obtaining energy and economic data was much more difficult than today, with so many databases freely accessible on the Internet. In this new context, we hope it will assist the readers in finding their way around the considerable amount of available information.

The "mind map" of Figure 1.1 summarizes the topics that are covered in this introduction to global energy issues and indicates the corresponding chapters.

The main elements of the problem can be stated as follows:

- we live in a world whose population is growing strongly, albeit in different ways in industrialized and developing countries;
- economic growth is considered worldwide as an imperative, which requires energy;
- energy is to be considered as an essential factor for development;

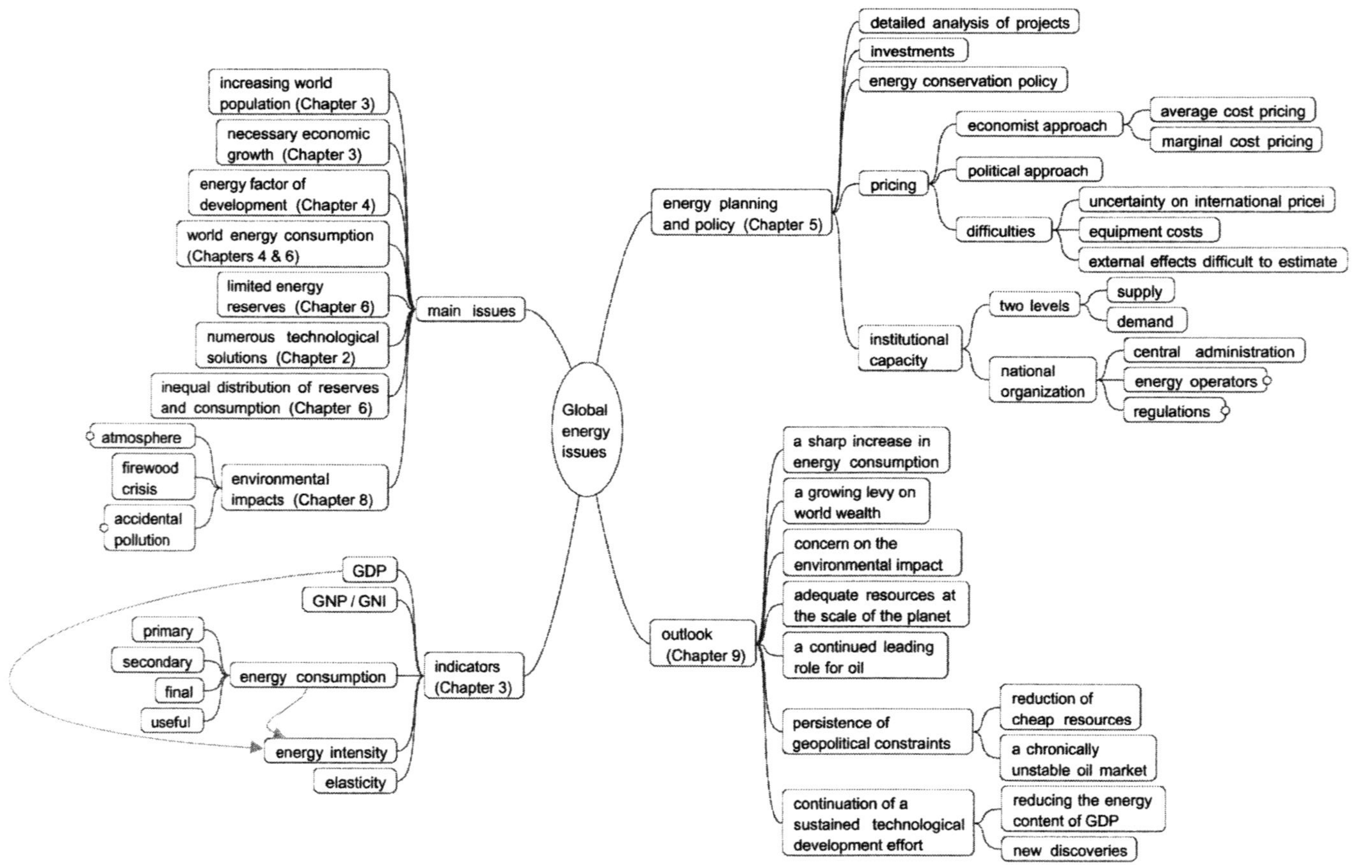

Figure 1.1 Energy issues mind map.

- energy consumption is thus expected to grow substantially;
- reserves and resources of fossil fuels are limited, and it will require at least several decades before renewable energies take up the baton;
- uneven distribution of energy reserves and consumption induces a trade that can be a source of tension;
- high levels of energy consumption have a significant environmental impact.

To quantify the relationship between energy and economic activity, indicators can be used. Their study is essential for energy planning and implementing informed energy policies regarding demand management, pricing and institutional choices.

This book presents these various topics in different chapters, before offering concluding perspectives:

- we can expect a sharp increase in energy consumption in the coming decades;
- its toll on global wealth will increase;
- in the short and medium terms, there is no risk of energy shortage, global resources being sufficient;
- the impact of energy on the environment will remain a concern;
- the energy content of the economy should continue to decline;
- oil will continue to play a leading role;
- geopolitical constraints are expected to persist;
- only the pursuit of a sustained technological development effort will reduce the disorders associated with all of these constraints.

1.2 STRUCTURE OF THE BOOK

Energy is a vast field that can be addressed from multiple angles. The approach proposed here is to start by providing the reader with technical bases on energy, and thus energy supply, before considering the demand, that is to say, its socio-and macroeconomic dimensions, then addressing global issues related to energy, and finally the main issues that arise in this area today.

This book summarizes the main issues related to energy and requires no special prior knowledge, whether of economics, engineering or international relations. It consists of nine chapters, including this introductory first chapter.

Chapter 2 introduces the main energy sectors (oil, natural gas, coal, synthetic hydrocarbons, nuclear power, renewable energy, thermal or pneumatic storage), i.e. how the main sources of energy can be exploited. At times quite technical, it is the longest chapter of the book, and can be skimmed through at a first reading.

Chapter 3 presents the main macroeconomic and energy indicators that are commonly used to assess the energy situation in a country. The concepts that are introduced there will be used consistently in other chapters; it is thus essential to understand their definitions and limitations well. Its last section deals with international energy markets.

The fourth chapter analyzes the impacts of energy at the macro level, including the links between economic activity and energy consumption.

The fifth chapter introduces the main principles generally accepted in the development of energy policy and planning, and then discusses the institutional aspects.

The sixth chapter focuses on geopolitics: current consumption of energy, energy reserves and resources worldwide, and specific problems faced by developing countries.

The seventh chapter is devoted to the study of the energy situations in eleven different countries, highlighting the contrasts between them, as they depend on the level of economic development, demography, natural resource endowments, etc. The list of countries includes high-income developed countries (France, the United Kingdom, the United States), emerging countries, the so-called "BRICS" (Brazil, Russia, India, China, South Africa), a North African oil exporting country (Algeria), a West African country (Côte d'Ivoire), and an Asian exporter of coal and natural gas (Indonesia). A section is then dedicated to the European energy policy.

The eighth chapter discusses emerging issues related to energy, in particular its relationship to the environment and the success of policies aimed at controlling demand.

The ninth and final chapter begins with a prospective study of various scenarios for the medium and long terms. The analyses presented in this book are then briefly summarized by outlining the main pending issues.

1.3 EDUCATIONAL USE OF THIS BOOK

Let us now look at how this book can be used as course material for lessons. The issues addressed in this book have indeed been taught since the early 1990s at the École des Mines de Paris (Mines ParisTech) and other engineering schools in France, in order to provide engineering students interested in energy with a general knowledge of the topic through the concepts discussed here.

At first, the teaching took the traditional form of lectures, more or less following the order of the chapters, but over time this form appeared not to be the most appropriate one.

1.3.1 Recommended pedagogy

A new formula was put in place from 2004 onwards: chair lectures were removed and replaced by online courses that students take independently. They then work in small groups (typically four to a group) in order to prepare an analysis of the energy situation in a country, with the aim of giving an oral presentation to their peers and delivering a small written report, typically ten or twenty pages long. Ideally, of course, one of the students in the group would be a citizen of the studied country.

Data is provided to them (national final energy consumption and energy production from IEA energy balances[1] and BP statistical review of world energy), while they must complement this information by drawing on other sources, including from the studied country itself.

1 An energy balance or energy mix is a table or a graph showing the breakdown by source of energy consumption in a country or region for one or more years.

They are asked to analyze a number of indicators of the chosen country and group in tabular form the changes in recent decades: energy consumption, CO_2 emissions, per capita consumption, energy and carbon intensities, per capita GDP in current and constant value. The amount of time devoted to this activity is about 12–18 hours.

In this form of teaching, students independently learn all the concepts they must master in order to prepare their case study. Two types of media are available to them for this purpose:

- the textbook, in paper format;
- online learning sessions with soundtrack (called Diapason, which stands for *Diaporamas Pédagogiques Animés et Sonorisés* in French), which introduce the main concepts.

The Diapason modules are educational animated slide shows, each provided with a soundtrack. They exemplify Information and Communication Technology (ICT) applied to education, to present theories, methodologies and technologies.

Annex 1 provides the reader with additional information on how to actually implement this pedagogy.

1.3.2 Available Diapason sessions

The list of available Diapason sessions[2] is given below:

- Sectors and global indicators:
  - energy sectors;
  - energy and economic indicators.
- Geopolitical aspects:
  - world energy consumption and production:
  - energy problems of developing countries;
  - reserves and resources;
  - energy markets.
- Recent changes:
  - environmental issues;
  - changing technology;
  - objectives and impact of energy policies.

1.4 ENERGY AND ECONOMIC STATISTICS

This book uses extensive data on energy consumption and production as well as macro-economic indicators. You will find in the Thermoptim-UNIT portal a page[3] containing links to organizations that identify and publish energy and economic statistics.

2 http://www.thermoptim.org/sections/logiciels/divers-portail/seances-disponibles
3 http://www.thermoptim.org/sections/problemes-globaux/statistiques

The energy and economic data we used in this book come from many sources. The principal sources are:

- the World Bank, which offers free online access to various comprehensive databases, including the World Development Indicators (WDI: http://databank.worldbank.org/data/home.aspx);
- the Population Division of the Department of Economic and Social Affairs of the United Nations, which collects very detailed information on world populations and provides it freely as spreadsheets (http://www.un.org/esa/population/);
- Energy statistics, published annually in June by BP: BP Statistical Review of World Energy, which include numerous data sets on oil, natural gas, coal, renewable energies, collected in a spreadsheet format (http://www.bp.com).

The IEA also publishes very detailed energy balances, but they are not freely accessible.

Generally, we use units of the International System or those directly derived from it in this book. Whenever possible, we also provide the corresponding values in the British system, for example by adding an additional right scale in the figures presented.

In addition, Annex 2 provides some general information on energy equivalences.

Chapter 2

Energy sectors

2.1 PRESENTATION

This chapter, which is the longest in the book, seeks to provide the reader with technical bases on energy: it briefly presents the energy supply, that is to say the different energy sectors. After describing the different forms energy takes and explaining the possible modes of conversion between them, it introduces the main characteristics of these sectors, starting with fossil fuels (oil, natural gas, coal, synthetic hydrocarbons), followed by nuclear, power generation and renewable energies (geothermal, solar, wind, hydro, marine energy) and finally presenting the thermal or pneumatic storage.

The reader interested in in-depth analyses of thermal conversion of energy may refer to [Gicquel, 2012], which provides a comprehensive presentation of engineering thermodynamics applied to energy systems.

This chapter ends with a section that, before comparing the various types of energy, mainly from an economic viewpoint, summarizes the main characteristics of the energy sector: long implementation times, diversity, rigidity, social acceptance questioning.

2.2 FORMS OF ENERGY

Energy can be defined as the faculty that a body has to provide mechanical work or its equivalent.

It comes in different forms (Figure 2.1):

- mechanical energy;
- heat;
- electromagnetic radiation;
- electricity;
- chemical energy;
- nuclear energy.

Mechanical energy comes in two main forms:

- potential energy, mainly derived from the Earth's gravity, is that which a body at rest located at altitude possesses: it is for example energy contained in the water of a dam, in the counterweight of old clocks, etc.
- kinetic energy is that of a body in motion: wind energy, water courses, crankshaft, flywheels, etc.

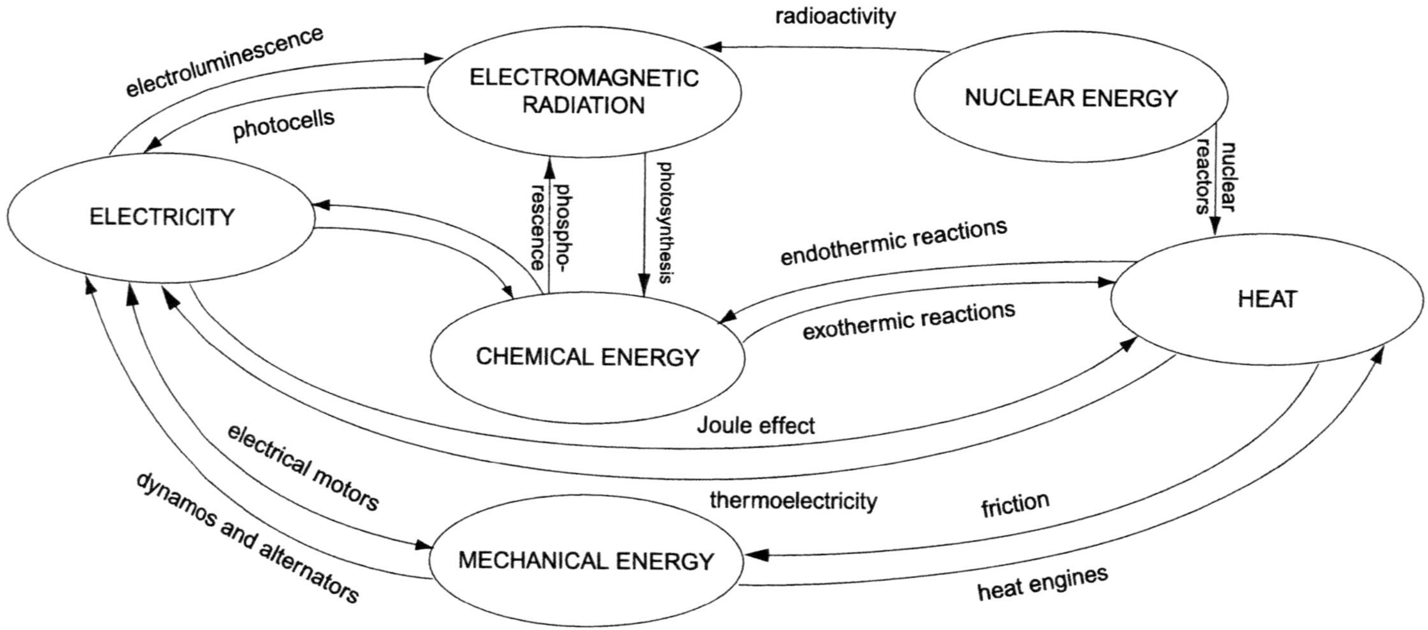

Figure 2.1 Different forms of energy.

Heat or thermal energy is energy of a body whose temperature is higher than that of its surroundings. There are two forms of heat: sensible heat, involved during a temperature change without a change of state, and latent heat, released (respectively absorbed) with no change in temperature during the solidification (resp. fusion) or the condensation (resp. boiling) of a substance.

Heat can be transmitted in three main modes of transfer: conduction and convection, acting by contact, and thermal radiation, which allows for remote transmission.

Electromagnetic radiation, which includes thermal radiation, is another form of energy capable of acting at a distance, even without a material medium.

Chemical energy is the binding energy between the molecules of a substance. Hydrocarbons (oil, gas, coal) are forms of chemical energy that play a fundamental role in the current global energy balance.

Electrical energy is a phenomenon produced by the movement of free electrons in conductors. It is a "noble" form of energy (see below) with three effects (thermal, magnetic and chemical) and a multitude of uses, some of which are specific: lighting, electric motors, electrolysis, induction heating, acquisition, processing and transmission of information, etc.

Nuclear energy is the binding energy between the particles that make up the nucleus of an atom. The theory of relativity showed the equivalence of matter itself and energy, according to the famous formula $e = mc^2$. Nuclear energy can be released by fission, which involves splitting the nucleus of a heavy atom (uranium, plutonium), or by fusion of light nuclei (hydrogen, deuterium). Today, only fission reactions are controlled. Fusion can as yet only be realized in atomic bombs, although research programs such as ITER aim at controlling it. A significant development of fusion is unlikely before 2050.

2.3 CONVERSION OF ENERGY

Between these different forms of energy, multiple conversions are possible (see Figure 2.1), these conversions being subject to two main laws of thermodynamics.

The first law or principle, of conservation of energy, states that in a closed system, the amount of energy subject to a process of transformation is fully incorporated in other forms when this process has ended.

If the first law states that all forms of energy are equivalent in value, the second law introduces a notion of quality: energy degrades from noble energy (mechanical, electrical) to heat. According to this law, the amount of usable (noble) energy must decrease in an isolated system, because of the existence of irreversibilities. In addition, the upper limit of the conversion efficiency of thermal energy into mechanical work is equal to the Carnot efficiency: $\eta = 1 - Tc/Th$, where Th is the temperature of the source of thermal energy (heat source), and Tc that of the ambient environment (cold source), both expressed in Kelvin or Rankine.

In practical terms, the main energy conversions are as follows:

- photosynthesis ensures the transformation of solar radiation into plant biomass and plankton which, over millennia, have given birth to all of the world's reserves of fossil fuels;

- the combustion of coal, oil and biomass can provide heat, used in many industrial processes, heating, etc. The amount of heat that a given fuel can provide is called its heating value;
- thermal machines allow for transforming either heat into mechanical energy (motor cycles), which can be used directly or converted into electrical energy, or mechanical energy into heat (refrigeration cycles), for producing cold for example;
- dynamos and alternators, driven by a source of mechanical energy (motor, turbine), produce electricity;
- friction and the Joule effect convert mechanical or electrical energy into heat;
- nuclear reactors transform fission reactions into heat, which can then be used to generate electricity.

Given that energy conversions are partly irreversible, the efficiency of these processes is lower in practice than the theoretical maximum efficiency.

During combustion the reaction products are found in the gaseous state, and it is possible, at a low temperature, that some of them become liquid or even solid, releasing heat of condensation or solidification.

It is notably the case for the combustion of hydrocarbons, water appearing among the products. The maximum energy release is obtained when the water contained in the flue gases is liquefied. The value of the heat of complete reaction is called higher heating value, or HHV. In the most general case where all the produced water remains in the vapor state, it is called lower heating value or LHV.

Table 2.1 shows the current and theoretical limits of some energy conversion modes.

2.4 MAIN SOURCES OF ENERGY

Nature provides two major categories of energy sources: non-renewable and renewable energies.

Table 2.1 Key energy conversions.

	Conversion of	*In*	*Best practice efficiency (%)*	*Best theoretical efficiency (%)*
Electricity production from combustible	Chemical energy	Electricity	32–42 (>50 CC)	85
Hydro power	Mechanical energy	Electricity	95–98	100
Fuel cell	Chemical energy	Electricity	50	80–100
Photovoltaic	Electromagnetic radiation	Electricity	20	50
Solar heating	Electromagnetic radiation	Heat	30–50	100
Internal combustion engine for transportation	Chemical energy	Mechanical energy	30–40	75–85
Photosynthesis	Electromagnetic radiation	Chemical energy	1–3	6

Non-renewable energies are mainly fossil fuels (oil, natural gas, coal), which are forms of chemical energy from biomass stored in past millennia, as well as the nuclear fission of uranium. As their name suggests, reserves for these energy sources are limited and do not replenish themselves, at least not in human time scales.

Their capital has already been largely tapped, so the baton must be taken up by other sources in the medium term.

Renewable energies were the first used by man, and largely continue to be used in traditional societies. This term covers a wide range of energy forms such as geothermal, solar, wind, hydro, and tidal energy, or energy of draft animals. Being in the form of flows instead of stock, their development has advantages but also poses specific problems that will be discussed later. However, they are practically inexhaustible.

2.4.1 Oil

Since the first oil production in 1859 by Drake in the U.S., the consumption of petroleum products has been subject to considerable development (1 GT in 1960, 2.9 GT in 1987, and 4 GT in 2007), due to the advantages presented by this energy: its liquid form, its high capacity energy density, and its relatively low environmental impact.

Liquid oil is easily transportable and can be distributed in variable amounts. Transportation networks and distribution of petroleum products, after departure from a refinery, are the least expensive and most flexible ones. In particular, oil nowadays is the form of energy most suitable for road, air and sea transport, all three being virtually captive markets for liquid fuels.

Another advantage offered by oil is the multiplicity of its uses. They have grown steadily over the last hundred years and now cover a huge range of products, as shown in Figure 2.2. The versatility of oil thus makes it a particularly attractive source of energy, and explains the strong growth of its market in the twentieth century.

Oil and natural gas, mixtures of carbon and hydrogen, or hydrocarbons, originate from microorganism plants and animals accumulated at the bottom of the sea, which, due to geological upheavals, have been transformed into sludge gradually filtered by

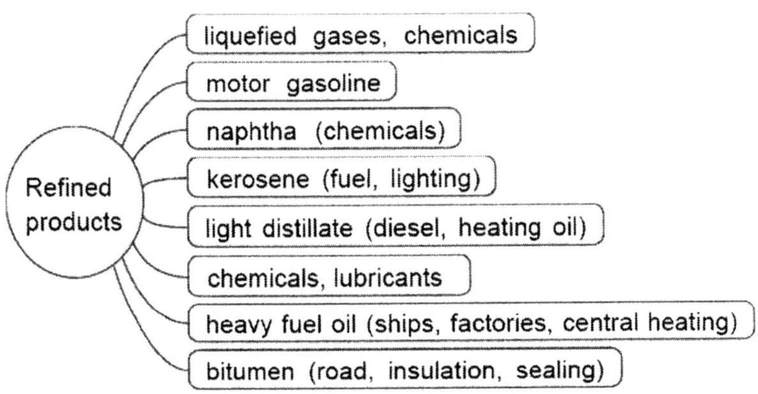

Figure 2.2 Different uses of oil.

porous sedimentary rocks, and moved so that, blocked by impermeable soil layers, they form deposits.

The deposits consist of very variable dimensions extending to depths ranging from surface outcrops to almost 5,000 m. Their geological composition corresponds to porous rocks in which hydrocarbons are trapped, their porosity (and hence the concentration of oil) being at best equal to 25%. These rocks are called reservoir rocks or rocks stores, as opposed to bedrock where hydrocarbons were formed originally.

Oils are mineral mixtures of varied compositions and qualities. They are often characterized by their density, expressed in degrees API according to the American Petroleum Institute. This degree varies between 80° API for lighter and less than 10° API for more viscous oils. Beyond 20° API, they are called conventional oils, or just oils; below, between 10 and 20°, heavy oils.

The degree API should not be confused with the physical density, which ranges from 0.7 (light oils) to about 1 (heavy oil). The relation between the degree API and the physical density is:

$$d_{API} = 141.5/d_{16°C} - 131.5$$

2.4.1.1 The oil sector

As shown in Figure 2.3, the oil industry has five main branches.

To extract oil deposits, one must drill wells whose unit cost is very high, especially when it comes to underwater (offshore) deposits. To limit the number of drillings, exploration has developed data mining methods from geology and geophysical knowledge.

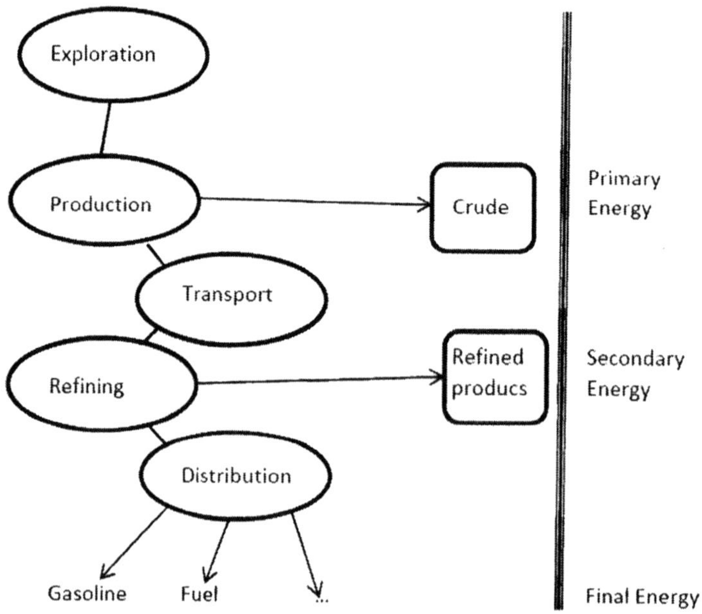

Figure 2.3 Oil sector.

Geological studies, based on the observation of the surface, possibly assisted by aerial photographs, help to identify the structure of the field and to locate areas with petroleum potential.

Geophysical prospecting is used to characterize the properties of a field, through gravity measurements (local variations of the gravity field), magnetometry (magnetic field variations), electrical resistance, and above all through seismic studies – consisting in following the path of artificially created underground shock waves (explosives or vibroseis). The computer analysis of the records now allows for fine interpretations.

When prospective studies suggest that a field exists, exploration drilling is undertaken at depths ranging generally between 800 and 5,000 m, but which may even exceed 10,000 m.

The execution time of a drill ranges from 15 days to several months, at a rate of a few meters per hour. On average, only 1 in 5 or 1 in 10 of the explored areas actually contains oil resources. When this is the case, the rig is removed and a characteristic set of valves called a "Christmas tree" remains on the ground. Otherwise, the hole is closed.

Most modern oil explorations, even based on state-of-the-art geological and geophysical techniques, remain highly uncertain and entail a high financial risk reserved for large companies. Moreover, exploration can only be conducted when these companies' profits are sufficient, i.e. when oil prices are high. The pace of discovery therefore significantly follows that of prices, adding yet another fluctuating factor to the supply/demand balance equation.

There are significant disparities in the pace of oil exploration around the world: for an average of 95 drilled wells per area of 10 000 km^2, the density of drilling exploration is 500 in North America, 90 in Eastern Europe (former USSR), 45 in Western Europe, 12 in Latin America, 7 in the Middle East, 5.5 in Asia/Pacific, and 4.5 in Africa.

These differences are related to many factors, including the likely success rate, best in relatively well-known zones, the ownership of the ground (which may belong to either the landowner or the State[1]), and, above all, the political risk in case of discovery. Many international companies, fearing nationalization, are reluctant to engage in exploration programs in some developing countries.

Finally, if the success rate of exploration tends to grow because of advances in survey methods, the size of the discovered fields tends to decrease. In order to measure the results of the exploration activity, a discovery index is defined: it is the amount of discovered oil divided by the exploration effort. It is expressed in barrels per drilled foot (b/ft). The curves representing the evolution of the discovery index clearly reflect a downward trend, except for the 1950–1970 period, when offshore exploration boomed. The values vary considerably depending on geographical area: in the United States, they are now particularly low, circa 20 b/ft, whereas the world average is around 500 b/ft.

This sharp reduction in the size of discoveries suggests that it is unlikely that many other "super-giant" fields comparable to those of the Middle East will be discovered in the future.

1 This is the most general case, the law of the United States being the exception.

In total, worldwide, between 30,000 and 35,000 petroleum basins have been discovered. Most are small, while only 1% of them contain 90% of world reserves. There are now about 600 high capacity tanks, among which 200 have never been explored and 240 have more or less been explored, but are not productive. Out of the 160 major productive ones, 25 have a capacity of more than 10 billion barrels (10 Gb), 6 a capacity greater than 50 Gb, and one contains more than 50% of all the world's oil. Among the 200 unexplored basins, it is estimated that 12 to 15 have a capacity greater than 10 Gb, but none would be able to compete with the Middle East.

2.4.1.2 Production

A deposit is exploited by drilling several wells. According to the natural pressure in the reservoir, production can occur spontaneously or requires the use of pumps (surface suction/pendulum pressure pumps, or, increasingly, electric pumps in the well).

These means correspond to the primary recovery, which generally does not collect more than 20% of the potential of the deposit.

To improve the efficiency, it is necessary to systematically inject water or gas under pressure in one or more locations, thereby facilitating the surge of the oil to production wells. This is called the secondary recovery. On average, it can improve the recovery rate to 33%.

Finally, the tertiary recovery phase, much more expensive and under ongoing technological development, is expected to increase the rate of extraction to 40 or 50% by injecting steam, miscible hydrocarbons or diluted chemicals in water before decanting the extracted oil to separate the gas and salt water.

These techniques are expensive and therefore only justified if prices exceed certain thresholds.

The complete exploitation of an oil field takes several years and can take up to 15 years for a large offshore field. In the latter case, it is necessary to use platforms that are set down or anchored to the seabed, depending on depth.

The development of offshore exploration and production was made necessary by the rapid depletion of the ground reserves. The oceans representing 70% of the Earth's surface, the potential oil they contain is indeed considerable, even if the environment is much more hostile.

Initially developed near the coast where large deposits were, offshore exploration and production then extended to access much more difficult areas such as the North Sea, the open sea off the coast of Labrador, etc. Approximately 30% of the oil produced in the world today comes from the seabed, and this proportion is expected to remain stable in the near future.

2.4.1.3 Transport to refineries

From production wells to refineries, crude oil transportation is provided either by tankers or by pipelines.

The advantage of delivering it by sea is that the destination of the cargo can change at any time, although tankers must return empty to places of loading. We will see later that this flexibility has enabled the development of a competitive market: the oil spot market.

Note that the international oil trade still accounts for a very significant share of world seaborne trade (33.6% in 2009 against more than 50% from 1950 to 1980 [UNCTAD, 2012]).

The main technical constraint is that crude oil can hardly be used as it is, and it is therefore necessary to refine it to produce "derivatives" or "products", which correspond to very different uses, either related to energy or chemistry.

2.4.1.4 Refining

Refining is separating the various hydrocarbon oil components. This is done using physical and physico-chemical processes that yield four major product categories or "cuts": Liquefied Petroleum Gas (LPG), gasoline, middle distillates (light fuel oil and diesel), and heavy fuel oil, as shown in Figure 2.4.

The atmospheric distillation unit allows for separating primary products at atmospheric pressure, depending on their volatility. Oil is introduced at 380 °C, and the vapors are then separated by condensation at temperatures ranging from 120 °C to 255 °C. At the bottom remains a heavy residue, called atmospheric residue.

To recover it, it is distilled under vacuum at 480 °C, which provides a distillate that catalytic cracking can separate into lighter products. More gasoline fuel is thus obtained.

Reforming is used to transform the heavy gasoline fuel to high octane fuel. It is obtained by passing the gas over a platinum catalyst at 500 °C.

Note that refining is a highly constrained problem, the products depending to a large extent on the composition of the incoming crude. Refineries are not able to

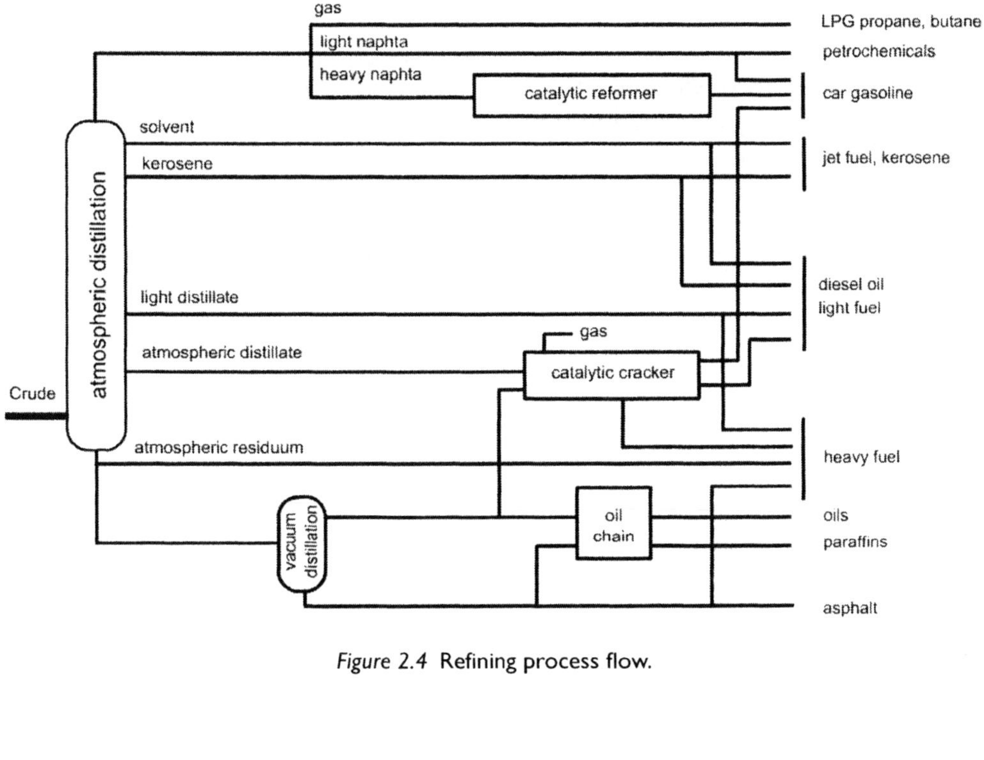

Figure 2.4 Refining process flow.

easily adjust their production to a strong evolution of demand, and a large part of the difficulty in this area stems from that rigidity.

Indeed, the growing trend to dedicate oil reserves to specific uses (road and air transport) has induced a significant reduction of the global demand for heavy products since the first oil shocks. To cope with this development, and to enhance the value of oil, deep conversion technologies are being developed in the most efficient refineries.

These very expensive technologies can "crack" heavy molecules to obtain lighter ones. It is thus possible to reduce the production of heavy cuts threefold compared to a refinery without a conversion unit, and by a factor of two compared to a refinery equipped with a conventional conversion unit.

Table 2.2 gives the effect of a standard conversion unit (catalytic cracking) on the production structure of a refinery.

In recent decades, the refining industry of oil-importing countries has faced a necessary restructuring for several reasons: because the decline in oil demand has resulted in losses of market shares, because of the evolution of the demand for finished products towards lighter cuts, and finally because of the relocation of refining plants in producing countries.

For many countries, all of these problems have led to the need to import more "white products" (gasoline, diesel). Globally, however, the trade in finished products only accounted for 30% of the total trade of oil in 2010 (crude and products) in volume and represented 20% of the consumption.

2.4.1.5 *Distribution*

While production and refining activities remain centralized, distribution of petroleum products may be more widely scattered.

The method of distribution depends on the demands of the consumer:

- large consumers generally receive their product directly from the refineries;
- final delivery to small consumers is only by road; e.g. in France 32,000 service stations supply the 21 million cars and road transport, and 6,000 dealers in oil supply more than 6 million central heating systems.

Oil companies must have deposits, used as strategic storage (90 days' consumption for the Member States of the European Union, and imports for those of IEA) and to respond to economic fluctuations in demand. They are usually located close to consumption areas.

Table 2.2 Structure of production of two refineries.

	Refinery without conversion	*Refinery with catalytic cracking*
Gas and LPG	3.5	5.3
Gasoline Fuel	15.2	23.3
Diesel and heating oil	35.4	36.4
Heavy fuel oil	42.5	29.1
Refinery Consumption	3.4	5.9

2.4.1.6 Structure of the technical costs

The cost of oil can be decomposed into four factors: production, transport, refining and distribution. To these technical costs, the oil revenue to the producer and the fiscal taxes to the importer must be added, which may represent a significant portion of the price paid by the final consumer.

In the Middle East, the production cost is very low, about \$1–7/b. In more difficult areas (e.g. offshore), it is \$15 to 30/b [IEA World Energy Outlook 2008].

The cost of transportation of oil remains low, either by pipeline or by sea. It is estimated at around \$1.5 to 2/b.

We saw earlier that a refinery produces a range of finished products. Although some techniques for determining the marginal cost of each product exist (duality theory in linear programming), refiners are generally reluctant to define allocation refining costs too rigidly.

It is difficult to disaggregate the costs of refining. Overall, however, it can be estimated that the cost of refining ranges between \$2.5 and 4.5/b crude oil according to the qualities and the degree of cracking.

The cost of distribution depends strongly on the network used to deliver the finished product.

In case of direct delivery to a large industrial consumer, the price distribution is approximately \$4/b (heavy fuel). For deliveries in small quantities, it is about \$5–10/b.

2.4.1.7 Investments in the oil industry

Exploration and production is a high-risk activity given the political risks (possible nationalization of companies working abroad) and financial uncertainty (see above). It is very expensive, especially for offshore operations. Budgets spent by oil companies fluctuate widely depending on the price of oil, which determines their profits. As an indication, global spending on exploration and production ranged between \$400 and 500 billion from 2008 to 2011 [Saniere et al., 2010]. Spending in the refining industry is much lower (slightly above \$25 billion during the same period).

Due to the global slowdown in demand over the past fifteen years, the share of the distribution sector in the oil sector investments fell by almost half.

2.4.1.8 Conclusion

The versatility of oil makes it a particularly attractive source of energy, which explains the strong growth of its market in the twentieth century. Easily storable and divisible at will, it is a preferred fuel for road and air transport.

Given its technical and economic characteristics, exploration and production of oil is reserved for highly integrated businesses enjoying a sufficient financial strength to cope with the associated risks. In contrast, the distribution of petroleum products can easily be achieved in a decentralized manner and does not require any particular network, which makes it a preferred energy source for countries with limited infrastructure, in particular for developing countries.

2.4.2 Natural gas

Natural gas, as explained before, has the same origin as oil. It is like a mixture of hydrocarbons whose main component (between 70 and 95%) is methane (CH_4), its density being 0.55 compared to air.

One of the main qualities of natural gas is its relative cleanliness. Burning it only releases water and carbon dioxide, in proportions 2 to 3 times less than coal and 1.5 to 2 times less than oil. The environmental impact of natural gas is much lower than that of coal or oil, with almost no contribution to the emission of oxides of sulfur and nitrogen.

2.4.2.1 *The gas sector*

As shown in Figure 2.5, the gas chain consists of five main branches.

The first two branches are similar to those of petroleum, as natural gas has been formed by the same process and is sometimes located in the same area as an oil field. In that case it is called associated gas, which can be either dissolved in the oil or separated. Until the first oil shock, the associated gas was burned at the site of production in flares located near oil wells, or reinjected for secondary recovery operations. It is now increasingly used as non-associated gas, or dry gas, when economically viable.

Natural gas presents stronger constraints than oil. While the latter can be transported and stored in a relatively simple way thanks to its liquid form, with a low specific volume, natural gas, in gaseous form under standard conditions, needs to be

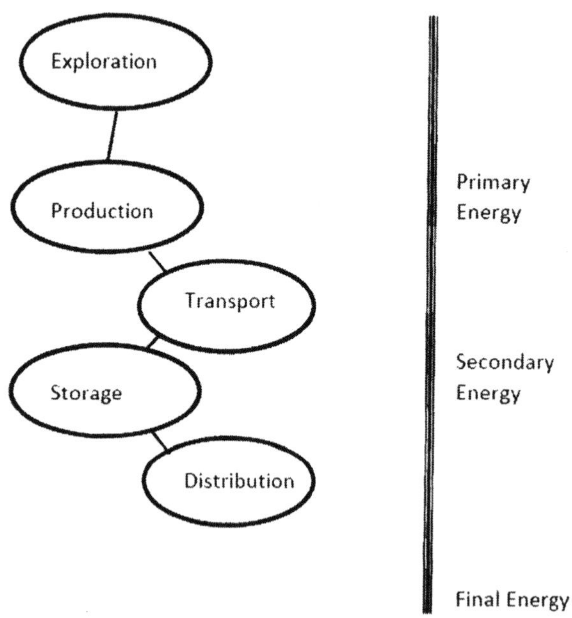

Figure 2.5 Natural gas sector.

packaged in a special way so as to be transported under safe conditions. In addition, the composition of natural gas varies considerably from one deposit to another and it often needs to be purified.

Transporting natural gas thousands of miles from the production site to the place of consumption requires the use of complex and expensive technologies.

Today, two main routes are used:

- transport by pipeline;
- transport by LNG (Liquefied Natural Gas) tankers.

With pipelines, the gas is transported under pressure in butt-welded steel tubular conduits with a diameter of 0.20 to 1.4 m, with recompression stations about every 80 km. Submarine pipelines have recently appeared, in addition to land pipelines. An interconnected network of increasing density has developed in recent decades, which has reached 550,000 km in the United States, 160,000 km in Russia, and 36,000 km in France. This heavy and very expensive infrastructure has the advantage that large flows can be routed in gaseous form. However, it is rigid and strongly links producers and consumers in the long term.

Figure 2.6 shows the transportation network of natural gas in the United States in 2009, which included 210 interconnected subnets.

LNG transport in liquid form also represents a technical feat. To liquefy the gas at atmospheric pressure, it must be cooled to −160 °C, which reduces its specific volume by a factor of 600.

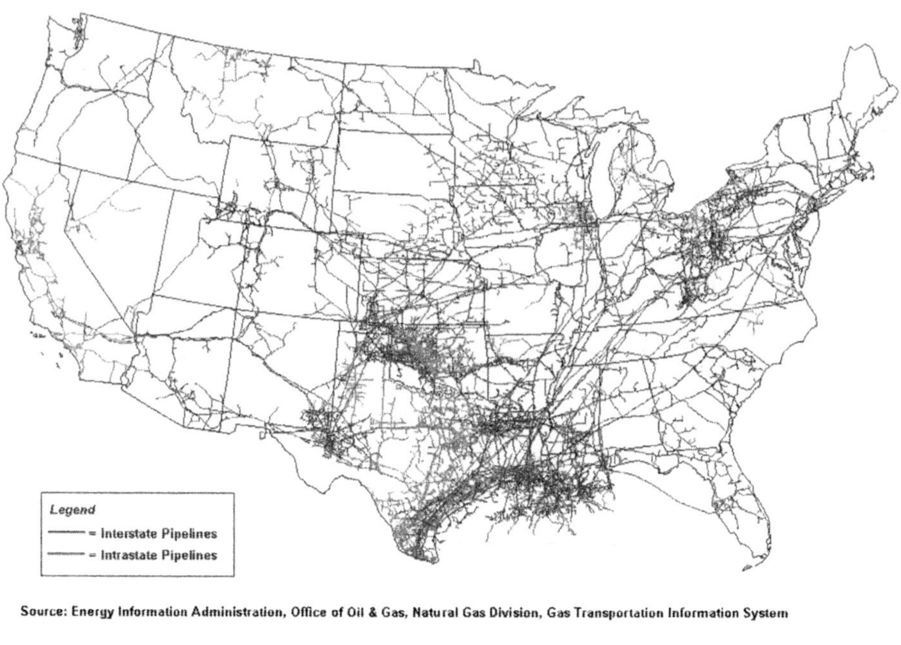

Figure 2.6 Natural gas transport network in the United States in 2009. [*Source*: U.S. Energy Information Administration (Sept 2012)].

LNG tankers with a capacity of 75,000 to 160,000 m^3 bring LNG to port.

The complete LNG chain includes:

- a pipeline, which connects the place of production to the port area;
- a liquefaction terminal and a storage unit of liquefied gas;
- a fleet of tankers;
- a regasification terminal;
- a network of pipelines to ensure gas distribution.

At the end of 2011, 89 LNG regasification terminals were in operation around the world, against 40 in 2001, as the number of importing countries increased during this period from 10 to 25. Japan and Korea alone accounted for 43% of global regasification capacity.

Note that the natural gas liquefaction consumes a significant amount of energy, and that losses occur at different stages of the LNG chain. The resulting consumption is about 13% [Babusiaux et al., 2002].

In addition to these transportation facilities, storage in aquifers or salt cavern is necessary to adapt supply to demand, which varies greatly, especially between winter and summer.

Gas consumption is indeed likely to fluctuate significantly and abruptly, the main cause of these fluctuations being the change in the demand for heating needs. A pipeline network or a LNG chain cannot adapt to the gas demand load variation between summer and winter, which varies in a ratio 1 to 4.

As the user cannot store natural gas, the distributor has to do so. Two techniques are mainly used for the storage of large quantities, including inter-seasonal storage:

- deep aquifer storage, between 400 m and 1200 m, which uses a configuration consisting of a favorable geological anticline or bowl whose upper part is formed of impermeable ground (clay). The technique is simple, the gas pressure driving the water, but it leads to the formation of an unrecoverable gas bubble;
- salt cavern storage, between 900 and 1400 m deep, for large capacity storage, is achieved by creating an artificial cavity in solid salt, capable of withstanding the mechanical load, by dissolving the salt through the circulation of freshwater. The gas is then injected or pumped as needed.

A third possibility is storing LNG in tanks buried near the regasification terminals. This technique, which requires continuous cooling of tanks to compensate for the heat loss, is limited to smaller quantities.

The distribution network consists of pipes of varied sizes (8 cm to 1 m), operating at medium pressure (50 mbar–4 bar) and low pressure (20 mbar).

2.4.2.2 *Structure of the technical costs*

The cost of the gas chain essentially comprises three parts: production, transportation and distribution, the latter two being relatively high. The distinction between the

FOB (Free On Board) and CIF (Cost Insurance Freight) costs is essential in the case of natural gas.

Production costs are essentially the same as those of oil we discussed section 2.4.1.6.

For both the LNG chain and pipeline transport, the amount of investment required to transport the gas results in high transport costs: in 1990 it was estimated that transport accounted for 40% of the cost of Algerian LNG in France and 50% in Japan. In the case of transport by the Siberia-France pipeline, this percentage was 35%. Since then, costs of the LNG chain have been cut by about 25%.

The cost of distribution is also very high, given the weight of the distribution network, especially for domestic purposes:

- 10–20% of the delivered price in the industry in France and the USA,
- 40–45% of the delivered price to households in France and the USA,
- more than 50% of the price, even for industrial use, in Japan.

2.4.2.3 *Investments*

The entire chain necessitates considerable investments, both for the producer and the consumer, and thus binds them in the long term. Indeed, LNG tankers can unload their cargo only in regasification terminals, unlike oil, which can be unloaded in almost any port.

For example, the investment costs of a LNG chain are:

- liquefaction of 5 Mt/year: 1.08 billion 1999 U.S. dollars;
- 135 000 m^3 LNG: $0.2 to 0.25 billion/unit;
- LNG terminal: $0.5 billion.

For such a project, the CIF cost of production of LNG is $3/MMBtu (GBtu), or about 16 $/boe for a distance of 6000 nautical miles [Babusiaux et al., 2002].

2.4.2.4 *Conclusion*

Natural gas is one of the cleanest highest-energy-content energy sources and can compete with oil in a significant number of energy uses, as well as feedstock. However, given that it is extremely volatile, each branch of the gas chain is characterized by very high investment costs, which makes the gas industry particularly centralized and highly rigid.

2.4.3 Coal

Since the industrial revolution, coal consumption has increased in considerable proportions. In 1950, the share of coal still accounted for 57% of global consumption of commercial primary energy, and then declined to 27% in 2002. Since that date, it has remained stable between 25% and 31%.

The importance of reserves and their distribution complementary to that of oil make it a serious potential competitor for the latter, the main obstacle to its use being its much greater impact on the environment.

Coal was formed from huge masses of plants mixed with minerals during the Carboniferous period (about 300 million years ago). These masses are composed of carbon, volatile matter (hydrogen), oil and ash.

There are three main types of coal (or solid fuels): peat, lignite and coal, including anthracite, which is a variety of superior quality.

- peat was formed in the Quaternary. Fibrous, lightweight, highly impregnated with water, it is a poor quality fuel, which is hardly traded internationally;
- lignite dates back to the Tertiary. Its fibrous structure reveals vestiges of branches and tree limbs. Though better than that of peat fuel, its efficiency is low. It is quite widely used, although there are also few international transactions;
- coal was formed in the Paleozoic era. Its heating value is much higher than that of lignite or peat. Coal includes various types of coals, which differ in their carbon content. Anthracite and lean coal are used as fuel, while the distillation of coal provides fat lean gas, tar and coke.

Table 2.3 compares these various fuels.

2.4.3.1 *The coal industry*

Contrary to oil and gas, one refers to the coal industry rather than the coal chain.

Two main categories of products come out of the mine: high quality coal is either directly introduced on the market as coal and pellets after screening and treatment, or sold to the coking plant, while low quality products are used as fuel for the production of electricity.

In coking plants, a number of synthetic products (ammonia, methanol, etc.) are made in addition to coke, as well as coke oven gas and recovered products (tar, benzole, ammonium sulfate, etc.).

Table 2.3 Heating values of various fuels.

Product	*Higher heating value (MJ/kg)*	*Carbon content (%)*
Dry wood	18–21	<30
Peat	12–16	<50
Charcoal	33	
Lignite	28–34	50–60
Sub-bitumin. coal	27–30	60–75
Coal		
Anthracite	33.5–35	93–97
Lean coal	35–36	90–93
Bituminous coal	32–37	75–90
Oil	41–48	
LPG	45–50	
Natural gas	55.5	
Hydrogen	143	

Coal mining takes place in mines, opencast or underground:

- open pit mines are operated in a manner similar to quarries. Recovery rates can be very high (90%, against 40% for underground mines) and production costs are generally lower than underground. In contrast, open-pit mines can have disastrous effects on the environment. This form of production accounts for 80% of the production in Canada, 60% in the USA, 50% in Australia, 33% in Russia;
- underground mines need at least two wells to allow for ventilation, necessary to remove gases and cooling the mine. Surface facilities are grouped around the wells. Various methods of exploitation exist, which the reader can look up in specialized books. Underground mines pose significant security problems, because of the risk of collapse, firedamp explosion (because of a flammable gas consisting mainly of methane) and flooding. Besides such accidents, working in underground mines can also cause lung diseases: silicosis and miner pneumoconiosis.

2.4.3.2 Coal processing

To meet user demand, the extracted coal undergoes a transformation over several steps.

It is first cleaned, sorted, and graded to be separated according to size and quality.

Abundant and inexpensive varieties, known as steam-coal, are used for the production of energy. The size of the energy production units is indeed such that they can adapt to various types of coal.

Additional transformations can be made to respond to specific requests or more stringent emission standards (see Chapter 8). Nowadays, coal is thus crushed as finely as possible so as to increase the area of contact with the combustion air and get higher combustion temperatures. In doing so, the reaction efficiency increases while the emission of pollutants (nitrogen oxides, unburnt hydrocarbons) decreases.

One of the most promising techniques is the fluidized bed combustion, where coal, finely ground, is burnt in suspension in a current of air flowing upwards.

Coking or carbonization produces coke, fairly pure carbon that is essential to the metallurgy of cast iron.

Since 1973, a number of recovery processes have been studied:

- production of clean and desulfurized coal, under development in the United States;
- development of a mixture of coal/water to be burnt in fuel boilers;
- coal liquefaction by hydrogenation to produce fuels (methanol) and raw materials for the chemical industry. The most used techniques were developed in Germany during the Second World War, as the Third Reich had an imperative need for fuel for its army, and could tap the coal deposits of the Ruhr region;
- in situ gasification of coal in order to convert resources directly in the ground that are not economically viable by conventional methods. To this end, incomplete combustion is achieved in deposits by injecting pressurized air. This produces a lean gas which can then be upgraded in various ways, either by direct combustion or for the production of synthetic hydrocarbons.

2.4.3.3 *Transport of coal*

In producing countries, coal is mainly transported by rail. The development of coal pipelines or coal slurry is under consideration: the coal would be crushed and suspended in water conveyed in a pipe, then decanted and filtered on arrival.

Three quarters of the international coal trade is carried by sea, either by cargo ships or by tankers.

2.4.3.4 *Structure of the technical costs of steam coal*

The technical cost structure of steam coal consists of three parts: production, transport, and distribution. As for natural gas, the significant part of the cost of transportation makes it necessary to specify whether FOB or CIF prices are used (see above).

Coal must be extracted by human labor. As a result, the cost of production is a significant part of the price of coal: in France or Japan, it accounts for 55–60% of the supply price and for more than 40% of the delivered price in the industry.

The cost of production ranges from $25 to $35 per tonne FOB depending on the considered mine.

The cost of transportation obviously depends on the delivery distance. It is roughly equal to 40–45% of supply in France or Japan (CIF), and to 30% of the delivered price for the industry sector, which is about $22 per metric ton.

The distribution cost depends on the size of the consumer and its distance from the place of unloading. In France and Japan, it represents about 25% of the delivered price in the industry, or about $19/t.

2.4.3.5 *Conclusion*

Unlike other fossil fuels, the cost structure of coal is strongly affected by labor and transportation costs.

However, the main characteristic of coal has now become the pollution it entails. With its image of a "dirty" energy source, the use of coal is now mostly limited to the production of electricity and combustion in boilers of significant unit size, where pollution control equipment can be economically considered.

Given the existence of large reserves of coal, it is certain that it will have an important role to play in the future global energy mix, and the most promising techniques seem to be gasification and liquefaction. Its revival is linked to the outcome of the current research on "clean coal".

2.4.4 Non-conventional fuels

Thanks to the gasification or liquefaction techniques, research is underway to develop coal deposits previously considered non-economically exploitable.

In more general terms, alternatives to conventional oil are increasingly being developed, by using resources that are not directly usable at low cost as raw materials.

Thus a wide variety of liquid and gaseous fossil fuels have emerged during the last decades, resulting in a great diversification of the fuel supply, previously dominated by oil and secondarily by natural gas.

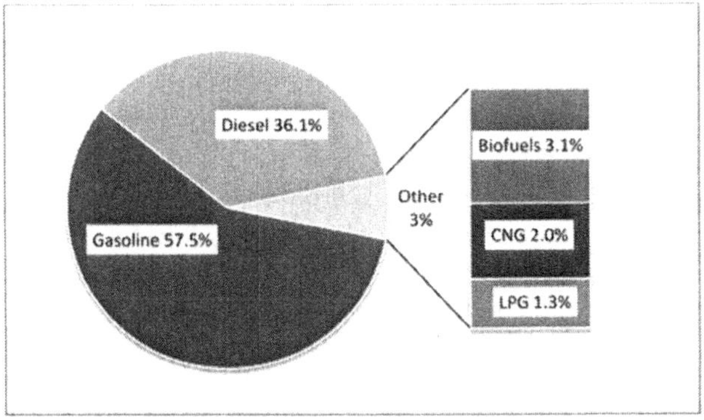

Figure 2.7 World distribution of fuels for transport in 2009. [*Source*: IFPEN].

The goal is to produce liquid or gaseous fuels in order to replace gasoline and diesel, which represent the quasi-incompressible demand of the transport sector (Figure 2.7) [IFP, 2009]. In this figure, CNG stands for compressed natural gas. These vehicles are fueled by natural gas contained in high pressure tanks (about 200 bar). The other liquid and gaseous fuels may be grouped under the name of non-conventional oils and synthetic hydrocarbons, or synfuels.

Experts have already begun to include some fuels once considered unconventional, such as shale, in the estimation of proved reserves and in the amount of oil produced. It is likely that this trend will continue in the future and that the boundary between oil and gas on the one hand and non-conventional fuels on the other hand will remain blurred and likely to evolve.

Note that this observation explains why the evaluation of oil and gas reserves has recently been significantly reconsidered.

It should be noted that the definitions of these fuels have varied widely over the past decades. Twenty years ago, the term synthesis hydrocarbon included all non-conventional production, but today this term is restricted to fuels produced by chemical means, and there is a preference to call unconventional oils those produced from shale sands or tar sands (see below).

For the sake of simplicity, we will group under the name of non-conventional fuels:

- oil produced from heavy oil and oil shale;
- biofuels, whether from the first or second generation;
- synfuels;
- liquid fuels from natural gas;
- methane hydrates.

2.4.4.1 Heavy oil and oil shale

The largest deposits of heavy oil and oil shale are located in Canada and Venezuela.

These include:

- Tar sands, crude oil of high viscosity that it is often impossible to extract the standard way;

- Oil shale, which is a sedimentary rock containing a solid organic material which can be extracted with heat in liquid or gaseous form.
- There are very significant identified reserves of oil shale (37% in Asia, 30% in the United States, 18% in the Commonwealth of Independent States (CIS), 10% in Latin America, mainly Brazil), tar sands (50% in Canada, the remainder in Venezuela, the United States and CIS), and heavy oil in Venezuela [Babusiaux, 2002].
- Although the percentage of waste can be very high, the exploitation of these reserves has begun, supported by high oil prices. Related technologies are just emerging and great uncertainties exist as to the investment and operating costs, which would be around 40 to 50 dollars per barrel of oil equivalent ($/boe). It should be noted, however, that as substitutes for fuels sold today at public prices are close to 220 $/boe, given the weight of taxation, synthetic hydrocarbons may be economically viable, especially when considering that, being locally produced, they have a positive macroeconomic impact on employment.
- However, their development may pose some serious pollution problems, as is the case for example for the oil shale industry: to produce one tonne of synthetic oil, 10 tonnes of shale are treated and 2.5 tonnes of water are used. In addition, the process rejects a waste volume 20% higher than the inputs (the so-called "popcorn" effect). The environmental impact of these technologies is far from negligible and appropriate solutions must be developed before production can be generalized.

Moreover, the exploitation of these resources may require large amounts of energy, particularly for injecting steam at a high temperature, so that their overall balance must be carefully estimated, particularly in terms of CO_2 emissions. Projects based on the use of nuclear reactors to provide the heat input are considered.

2.4.4.2 Biofuels

These are fuels produced from biomass. There are generally two types of biofuels: the first and second generations. Even though their definitions differ among authors, the first usually refers to biofuels that come into direct competition with food crops, while the latter are derived from lignocellulosic materials.

Some countries like Brazil and the United States have launched extensive programs to produce ethanol from sugar crops such as beet or sugar cane. Up to 10% ethanol can be incorporated in fuels for conventional gasoline engines, but its use as pure fuel implies significant changes in the fuel system.

In Brazil, the Proalcool program aims at substituting ethanol from sugar cane for gasoline. The development since 1975 of a refining (distillery) and distribution infrastructure has allowed Brazil, since 1984, to replace approximately 5 million m^3 of gasoline with ethanol, and convert 13% of the fleet to this national resource. Twenty years later, 80% of the fleet is "flex-fuel", i.e. it allows the consumer to choose, depending on prices, the cheaper fuel. In 2005, the Brazilian ethanol production amounted to 15 billion liters (15 million m^3), the cost of a liter of bioethanol ranging between $0.25 and $0.3 ($36–43/boe).

Biofuel for diesel engines is also produced from rapeseed oil or sunflower oil in Europe and the United States (between 7% and 30% can be added to diesel). This is known as biodiesel.

As of now, the main drawback of these programs is that they divert arable soil from food production, which may represent a high social cost, with the risk of causing famines in countries that implement them or even globally, a decline in the supply of agricultural products resulting in an increase of their cost.

2.4.4.3 *Synfuels*

Coal gasification or liquefaction processes are relatively old: in 1943, Germany produced up to 4 million tons of synthetic fuel.

Today, operational processes lead to prices ranging from $50 to $150/boe, but at the cost of high investments: $70,000 to 100,000/boe/d.

Methanol and synthesis gas could also be produced from coal, by combining electrolytic processes and gasification with oxygen.

To limit CO_2 emissions, electricity could be provided by a nuclear plant, and methanol and synthesis gas would be used to power gas turbines for peak periods. This way, a power generation fleet could be supplied from only nuclear energy and coal.

2.4.4.4 *Liquid fuels from natural gas*

These products stem directly from natural gas: they consist of ethane, propane, normal butane, isobutane and natural gasoline/condensate.

Blended with natural gas, they are extracted during its processing, before being transported to the place of consumption, and are substitutes for products refined from oil, their characteristic being their liquid rather than gaseous form.

2.4.4.5 *Methane hydrates*

Methane hydrates consist of water molecules trapping gas molecules that can store considerable amounts of gas such as methane (e.g. 164 cm^3 of methane in 1 cm^3 hydrate).

They take the form of flammable ice blocks found in sediments at low temperature and high pressure (in oceans or circumpolar regions).

Their development is under consideration, but still poses many technological challenges and could have a significant environmental impact.

2.4.5 Nuclear energy

Nuclear energy has been used since the mid-twentieth century, mainly for electricity generation and marine propulsion. It has undeniable advantages but also entails high risks, making it controversial.

Among its strengths are its high energy density (a 7 g pellet of nuclear fuel contains as much energy as one tonne of coal) and almost inexhaustible resource potential with certain types of reactors, the absence of greenhouse gas emissions, a low cost, and broad independency from geopolitical tensions.

Risks, in turn, are related to the hazards of uranium and radioactivity of the fuel and fission products, which make it mandatory to have draconian security measures in place throughout the fuel cycle and to use long-term waste storage. There is also a risk of proliferation of nuclear weapons with the currently operating reactors that produce plutonium.

2.4.5.1 *Fast and slow neutrons*

There are two main categories of reactors, depending on the level of neutron energy they use: fast-neutron reactors and slow-neutron reactors.

- "fast neutrons" are produced by fission reactions before they are slowed by a large number of shocks. Their energy is about 0.1 MeV to 2 or 3 MeV. These neutrons are able to split, thus destroying not only the nuclei known as fissile but also the actinides, which are heavier than the uranium nuclei accumulating in the reactor fuel. To burn this radioactive waste effectively, fast neutrons are required. It is said that the fast reactor operates as breeders, which means it produces more plutonium 239 (^{239}U) fuel than it consumes: fuel becomes inexhaustible;
- "slow neutrons" or "thermal neutrons" are slowed by a large number of shocks, usually in a medium called a moderator. Their energy's order of magnitude is one electron volt or a fraction of an electron volt, i.e. 6 orders of magnitude smaller than that of fast neutrons. They can only split a small number of nuclei: uranium 235 (^{235}U, the only uranium variety existing in nature), plutonium 239 (^{239}U) and uranium 233 (^{233}U) produced in the reactors. The global existing fleet comprises almost only slow neutron reactors.

"Epithermal neutrons" are neutrons with an energy ranging between the two previous modes. Such neutrons appear in molten salt reactors.

2.4.5.2 *Different types of nuclear reactors*

The notion of generations of reactors has been introduced to distinguish the main developments that marked the history of civilian nuclear reactors:

- the first generation corresponds to the first prototypes built mainly in the U.S. until the late 1950s. It encompasses the development of natural uranium graphite gas systems (*uranium naturel graphite gaz* UNGG in French) in France, with the completion of 9 reactors, whose efficiency was close to 29%;
- the second generation, from 1960 to 1995, marks the first phase of commercialization of nuclear power plants, with three main systems: light water reactors (BWR or boiling water reactors), pressurized water reactors (PWR, performance about 33%), and heavy water reactors (in Canada). Most operational plants currently belong to the second generation;
- the main objective of the third generation reactors is to increase plant safety, accidents at Three Mile Island in the United States, Chernobyl in Ukraine (formerly in the USSR) and Fukushima in Japan having shown the major risks associated with certain second generation units. The EPR (European Pressurized Reactor) reactor is a third generation reactor. Its performance, at 35%, is slightly higher than those of second generation;

- it is primarily the increased efficiencies which justify the current work on the fourth generation (Generation IV) fast-neutron reactors, their great innovation furthermore residing in designing the reactor along with the cycle that goes with it and optimizing this system as a whole, so as to properly tackle the issues of sustainability. The efficiency of PWRs, which are the most widely used reactors nowadays is indeed less than 35% in practice, while those of combined cycle gas turbines exceed 55%.

Since the early 1990s, different types of fourth generation reactor projects have emerged, with the objectives of blowing away technological barriers faced by earlier generations and achieving sustainable economic and technical performance, all with a better adaptation to sustainable development criteria.

Some of these projects involve high temperature gas-cooled reactors, called PBMR (Pebble Bed Modular Reactor) and GT-MHR (Gas Turbine Modular Helium-cooled Reactor). Some started in 2001, following the initiative of the United States' Department of Energy (DOE) called *Generation IV Nuclear Energy Systems Initiative*, which has managed to unite around it a group of nuclear research organizations from 10 countries in a loose cooperation called the GIF (*Generation IV International Forum*). Its goal is to study systems that could be the fourth generation of reactors, after those in use or planned in the short term (EPR), and deployable by 2030.

After studying in detail a hundred types of reactors, in September 2002 the GIF selected six concepts on which the efforts of its members will focus:

- Supercritical Water Reactor (SCWR);
- High-Temperature Reactors (HTR);
- Gas-cooled Fast-neutron Reactors (GFR);
- Sodium-cooled Fast-neutron Reactors (SFR);
- Lead-cooled Fast-neutron Reactors (LFR);
- Molten Salt Reactors (MSR).

2.4.5.3 Nuclear Fuel Cycle

By analogy with the other forms of energy that we have considered so far, we will call "nuclear fuel" the uranium producing slow neutrons that are further slowed down to provide heat at high temperature in the vast majority of reactors existing in the world, i.e. pressurized water or boiling water reactors.

In these reactors, as well as in those that the GIF is studying, the fuel is in the form of enriched uranium aggregates, as natural uranium lacks sufficient amounts of isotope 235.

The fuel cycle (Figure 2.8) is divided into three main phases: upstream, inside the reactor and downstream.

The upstream part includes the following main steps:

- mining;
- concentration and refining;
- enrichment;
- preparation of assemblies.

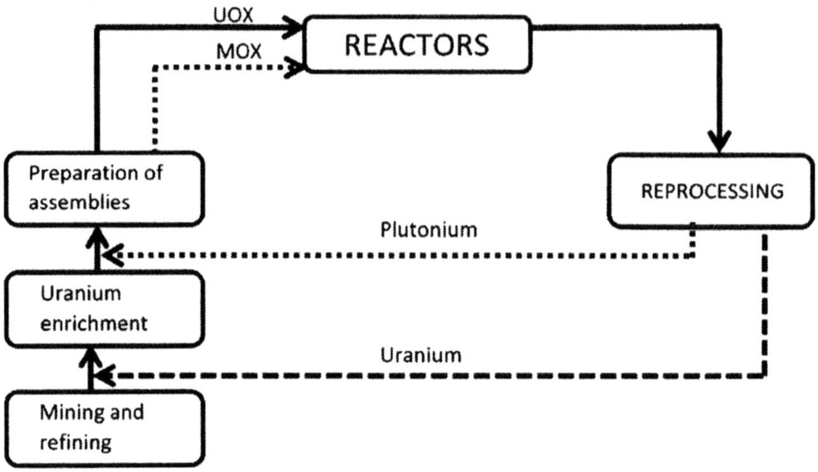

Figure 2.8 Nuclear Fuel Cycle. [*Source:* CEA].

The fuel is then used in the reactor, from which it comes out highly radioactive. Downstream includes the following main steps:

- reprocessing;
- waste packaging;
- waste storage.

In its natural state, uranium is a metal present in various concentrations in rocks such as pitchblende. Like coal, the ore is mined in underground or open pit mines. Its uranium content being very low (usually circa 0.1% to 0.5%), it is necessary to first concentrate it, in an operation realized at the site of extraction by chemical means, which leads to the production of a yellow paste called "yellowcake". Uranium concentration in octaoxyde of triuranium (U_3O_8) is 75%.

This yellowcake has to be refined to be purified and converted into uranium tetrafluoride (UF_4) or hexafluoride (UF_6).

It is then necessary to "enrich" the uranium so that the concentration in the isotope 235 is sufficient for the fission reaction to be maintained, since only this isotope produces slow neutrons. This requires increasing its ^{235}U concentration from a natural 0.7% to 3–5%.

Enrichment is very difficult to achieve because the technological properties of the two isotopes are very close. Their masses being slightly different, separation techniques are based on this property. Separation work units or SWU is the work required to separate a kilogram of uranium into two batches of different isotopic characters. Two methods are mainly used:

- gaseous diffusion, which is a gradual process, requiring many steps (1,400) and involving a large amount of energy (2,400 kWh/SWU);
- ultracentrifugation, at about 60,000 r/min, which requires much less energy.

Once the enrichment is successful, uranium, in the form of uranium hexafluoride (UF_6), is converted to uranium oxide, in the form of black powder, which is compacted into pellets.

These tablets are packaged in assemblies that form the heart of the reactor and whose 3–5% ^{235}U produce slow neutrons that sustain the nuclear reaction. Most of the uranium, ^{238}U, remains stable, with the exception of some nuclei that capture a neutron and become plutonium (^{239}U), also fissile as ^{235}U.

The fuel used in the reactor becomes radioactive and gradually degrades, mainly because of the decrease in the concentration of ^{235}U and production of fission products, plutonium and minor actinides. It is considered worn out when its composition is 95% uranium, 0.9% ^{235}U, 4% fission products, 1% plutonium and 0.1% minor actinides.

It must then be reprocessed, but this is only possible when it has cooled enough, which requires leaving it in a "pool" for three years.

Reprocessing serves a double purpose:

- gives value to the uranium and plutonium present in the fuel by recycling them, since they contain a lot of energy;
- extracts the non-recyclable waste by reducing and confining it.

Combustible materials are then recycled while waste is packaged for storage. We will not elaborate here on the techniques that are implemented for this, referring the reader to specialized literature.

All these operations are called the fuel cycle.

2.4.5.4 *Costs of the fuel cycle*

To obtain 1 kg of enriched uranium fuel, it takes about 8 kg of ore, costing $125/kg or $1,000 in total.

The conversion of uranium into UF_6 costs $12/kg of ore, hence about $100 in total for 1 kg of output [WNA, 2012].

The enrichment of one kg of fuel to 3.5% ^{235}U requires 7.3 SWU at $165/SWU, or about $1,200 in total.

The total cost of preparation of enriched fuel is about $2,300/kg of enriched fuel, to which must be added about $240 for the packaging itself.

Used in a 33% efficiency plant (second generation PWR) this fuel produces approximately 360 MWh of electricity, which corresponds to a cost of $7/MWh.

Of course this cost includes neither the amortization of the nuclear power plant, nor the cost of labor.

2.4.6 Renewable energies

Presumably promised in the medium and long terms to be significantly developed, most solar energy conversion technologies are not mature enough yet to compete with conventional energy on a large scale. There is, of course, an exception: hydro energy, while some other forms of renewable energy are beginning to be economically viable.

Technical constraints are mainly threefold:

- the available power density is relatively low, which entails large areas of collectors and high material costs;
- the variability of the source is high: solar energy fluctuates a lot, which often requires complex control systems;
- the need to store: being an energy-flow, it requires storage for most applications, which poses a problem because we do not presently know how to store energy in good conditions.

Economic constraints are twofold:

- a high capital cost: even when properly mastered the technologies are relatively expensive in investment, while their operating costs are generally low;
- the need for a back-up. In case of unavailability of the source, another energy is often required, imposing additional costs that can become significant.

Their development requires to control their complexity and to find appropriate financing mechanisms, which may be difficult to implement at a decentralized level.

Depending on their origin, we can distinguish five major categories of renewable energy:

- geothermal energy, provided by the Earth;
- direct solar energy (thermal and photovoltaic);
- indirect solar energy (wind, hydro, biomass);
- energy of the sea;
- energy from draft animals.

2.4.6.1 Geothermal energy

Geothermal energy comes from the gradual temperature increase as one penetrates deeper into the earth's crust:

- either because of the natural gradient (3 °C/100 m, with an average flow of 60 mW/m^2). In favorable areas one can find, at depths of 1,000 to 2,000 m or more, hot water reservoirs at temperatures of 50 °C to 90 °C, which can be used for heating. This is called low temperature geothermal energy;
- or because of geophysical singularities (high temperature natural geothermal reservoirs of porous rock). This is known as high energy geothermal heat. Installed capacity remains limited and highly variable depending on the country (2200 MW in the United States, 900 MW in the Philippines, 650 MW in Mexico, 500 MW in Italy, etc.).

More specifically, it is customary to distinguish three broad categories of reservoirs, according to their temperature levels:

- high temperature (>220 °C);
- intermediate temperature (100–200 °C);
- low temperature (50–100 °C).

In the first case, the geothermal fluid usually is essentially composed of water or steam, in the other two it is water, optionally under pressure. A special feature of geothermal fluids is that they are never pure water: they also include many impurities, corrosive salts (the concentration limit for an operation to be possible being 1.5 mol/kg) and Non-Condensable Gas (NCG) in varying amounts (0.1–10%). We shall see that this feature imposes constraints on the thermodynamic cycles that can be implemented.

For environmental reasons, the geothermal fluid should generally be reinjected into the reservoir after use, but this is not always the case.

The thermodynamic conversion of geothermal energy uses four main techniques:

- "direct-steam" plants can be used if the geothermal fluid is a superheated steam that can be directly expanded in a turbine. Historically, this type of plant was first implemented in Larderello, Italy in 1904;
- flash vaporization power plants can use sites where geothermal fluid is present in the form of pressurized liquid or liquid-vapor mixture. This type of plant is the most widely use nowadays. The geothermal fluid is expanded in a chamber at pressure lower than that of the well, thereby vaporizing a portion, which is then expanded in a turbine;
- systems known as "binary" use a secondary working fluid, which follows a closed Rankine cycle, the boiler being a heat exchanger with the geothermal fluid;
- fluid mixture systems, such as the Kalina cycle, are a variant of binary systems where the working fluid is no longer pure but consists of two fluids in order to achieve a temperature glide[2] during the vaporization.

Mixed or combined cycles can use both a direct or flash system and a binary system.

In 2009, the global installed capacity of geothermal electricity generation was 10.7 GW, with a production of approximately 67 TWh. Production costs range between $50/MWh (flash plants) and $110/MWh (central binary [IEA Technology Roadmap, 2011]).

2.4.6.2 *Solar energy*

Solar energy comes from thermonuclear reactions that occur within the sun, causing the emission of high power electromagnetic radiation, the sun acting much like a "blackbody" at 5,800 K.

Outside the atmosphere, the radiation received by the Earth ranges, depending on the time of the year, between 1,350 and 1,450 W/m^2. It is then partially reflected and absorbed by the atmosphere, so that the radiation received at ground level comprises a direct part and a diffuse part, the total ranging from 200 W/m^2 (overcast) to about 1,000 W/m^2 (zenith clear sky). The amount of energy received by a given surface depends on the local climatic conditions as well as on the device's tilt and orientation. Atlases of solar radiation at the ground level are issued by national and international meteorological services in the form of maps and charts, on paper or digitally.

2 The « temperature glide » is the technical wording to describe a temperature variation which occurs during the vaporization of a fluid blend.

Direct conversion of solar radiation is achieved in three main ways:

- in the form of heat;
- by photoelectric effect;
- by photosynthesis.

The thermal conversion of solar energy consists in intercepting the incident photons on an absorbing material, whose temperature rises.

Several modes of capture are possible:

- passive solar housing. For heating and space cooling applications, it is possible to design the architecture of the buildings so that they naturally (or "passively") optimize the use of the solar resource, without using auxiliary devices. The advantage of such a passive solar design of buildings is that it can lead to substantial energy savings with low incremental costs;
- flat plate collectors generally use the greenhouse effect to minimize heat loss from the absorber. Indeed, glass is transparent to visible radiation, and therefore lets the incident solar energy through, while it is opaque to infrared radiation, which has the effect of trapping heat. Depending on the technology used, the operating temperatures of the flat plate vary from 40 °C to 120 °C (vacuum collectors). Figure 2.9 shows the sectional view of a collector. The absorber consists of a metal plate on which the thermal fluid circulates inside welded pipes. Heat losses to the front of the collector are reduced by one or more glazings (2 on the figure) and those to the rear by an insulator;
- concentration collectors. To reach temperatures above about 120 °C, it is necessary to concentrate the solar radiations by using suitable sets of reflective elements (mirrors) or lenses (usually of the Fresnel type). The main constraint, apart from the higher cost of the devices, is the tracking system intended to follow the trajectory of the sun. A series of concentrators has been proposed and developed;

To thermodynamically convert solar radiation into electricity with a good efficiency, it is necessary to concentrate it, otherwise the heat losses of absorptive surfaces are very high and the collector effectiveness is too low. However, the costs of this industry and the technological difficulties encountered today limit its scope.

It is customary to characterize the concentration collectors by their concentration factor C, ratio of the surface of the collector and the one receiving the concentrated solar flux.

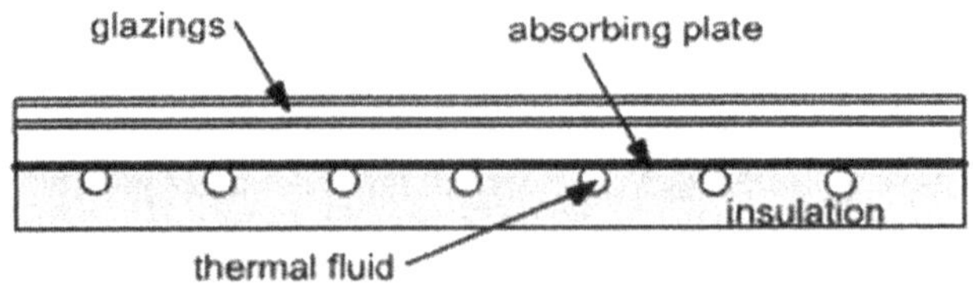

Figure 2.9 Collector sectional plane.

The experience of the last thirty years shows that four main technologies are actually used to achieve the concentration of solar radiation in technically and economically viable conditions:

- parabolic troughs;
- Fresnel lens concentrators;
- parabolic dishes;
- tower plants.

Figures 2.10 to 2.13 show the operating principles of these solar collectors. The first technologies only require to track the sun in one direction, but the concentration, and thus the collector temperature are lower (400 °C). The other two require a double tracking movement, but can reach much higher temperatures (750–1000 °C).

The parabolic trough ($C \approx 40$–80, Figure 2.10) comprises cylinders parabolic in cross section that concentrate sunlight onto a straight tube.

Fresnel linear concentrators ($C \approx 30$, Figure 2.11) use narrow rectangular plane mirrors for concentrating sunlight onto an absorber consisting of a fixed number of parallel tubes.

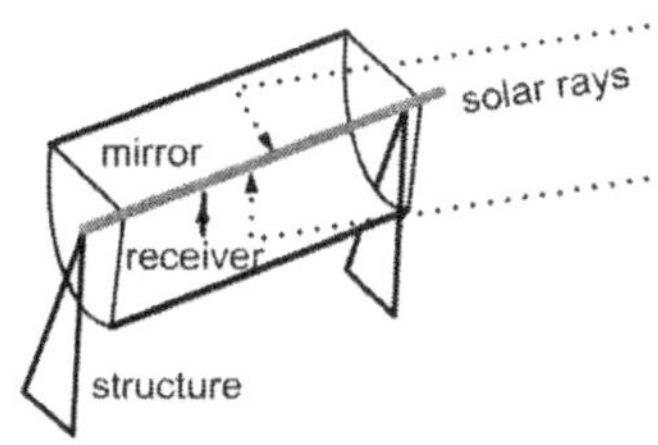

Figure 2.10 Parabolic trough.

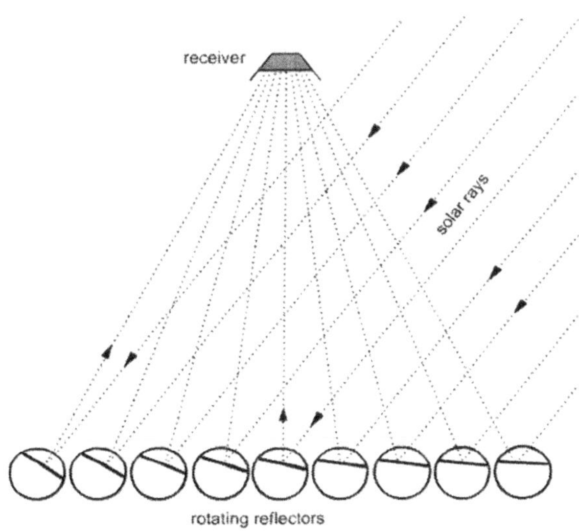

Figure 2.11 Fresnel lens concentrator.

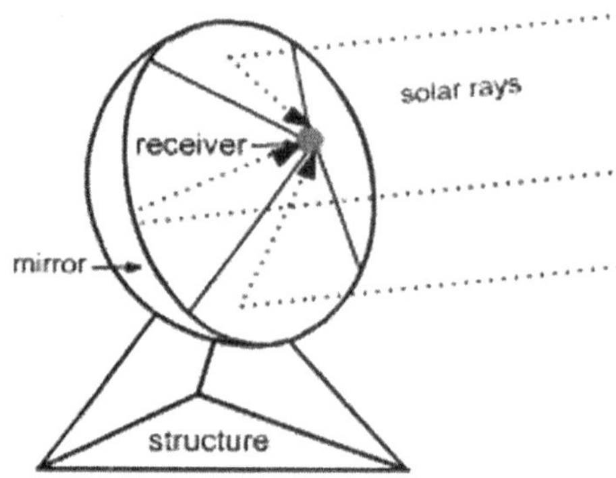

Figure 2.12 Dish collector.

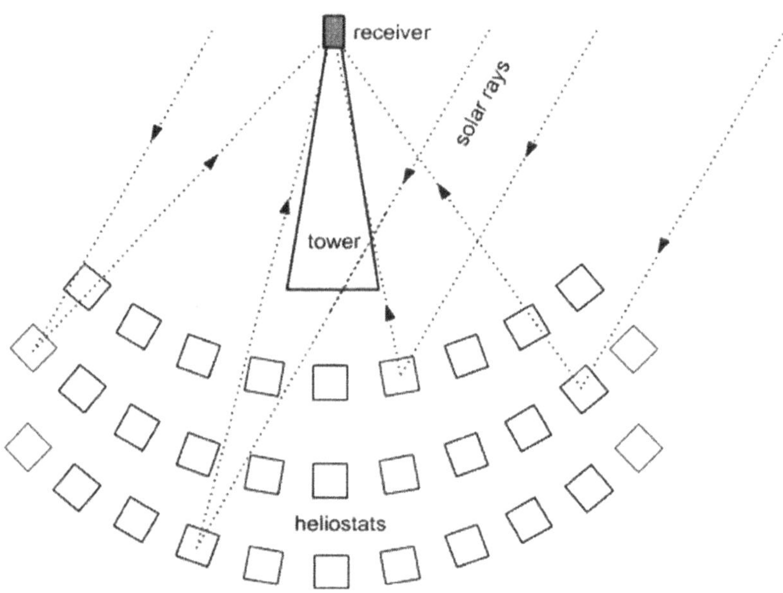

Figure 2.13 Power tower.

In the parabolic dishes ($C \approx$ 1,000–2,500, Figure 2.12), the reflector is a paraboloid of revolution.[3]

In the power towers or central tower plants ($C \approx$ 200–700, Figure 2.13), thousands of tracking reflectors, called heliostats, redirect incoming solar radiation to an absorber at the top of a tower, allowing to achieve both high concentrations and high radiation fluxes.

3 A paraboloid of revolution is the surface generated by a parabole rotating along its main axis.

Photovoltaic (PV) electricity is obtained thanks to doped semiconductors that, under the effect of the incident photons, produce a DC current, photon of sufficient energy hv allowing electrons to jump an excitation level (see Figure 2.14).

Originally developed to power satellites, solar cells are nowadays increasingly used for terrestrial applications, from solar calculators to high power generators.

Their main attraction is their simplicity of use: subject to direct or diffused sunlight, they produce electricity directly without moving parts, and require only simple maintenance.

The material used to make them is mostly silicon, which, depending on its crystalline state, leads to variable efficiencies (20% for the single crystal, 15% for the polycrystal, 8% for amorphous silicon, while the maximum theoretical efficiency is 43.5%). Other materials such as gallium arsenide are under study, and steady progress is being made year after year, foreshadowing significant improvements in performance in the short term.

In 2008, the global installed capacity of PV was approximately 14 GW peak, with an annual growth rate of 40%. Prices vary between $4,000/kW installed for large plants of several MW, and $6,000/kW installed for small installations [IEA Technology Roadmap, 2010].

2.4.6.3 *Biomass energy*

Photosynthesis is the property possessed by plants, through the action of chlorophyll, to convert electromagnetic radiation into chemical energy, stored mainly in themselves as carbohydrate, i.e. as sugar, and used for their growth and reproduction.

Under the action of sunlight, carbon dioxide from the air and water from the soil are processed as carbohydrates, from which all the components of the organic matter derive.

The maximum efficiency of this bioconversion does not exceed 6%, but considering the area covered by plants, this represents considerable amounts of energy.

The biomass comprises all organic materials, mainly of vegetable origin, natural or cultivated, on land or in the sea, produced by chlorophyll conversion of solar energy, fossil fuels excluded. The latter, as we have seen, are the byproduct of fossil plant materials corresponding to a form of biomass stored for millions of years.

Biomass produces three types of resources: forestry, agriculture and aquatic resources.

According to IEA, 10% of the global energy consumption comes from biomass, which makes it the most used renewable energy source.

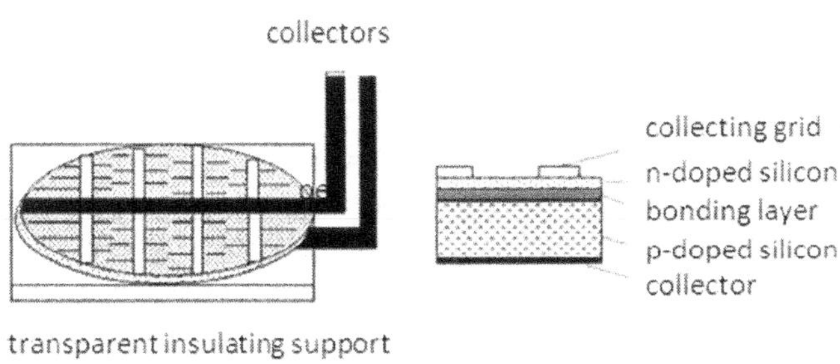

Figure 2.14 Schematic of a solar cell.

The biomass is mainly composed of lignin ($C_{40}H_{44}O_6$) (25%) and carbohydrates Cn(H_2O)m (cellulose $C_6H_{10}O_5$ and hemicellulose) (75%).

Biomass energy can be used in three main categories of processes:

- biochemical conversion: digestion, hydrolysis and fermentation;
- chemical conversion (esterification);
- thermochemical conversion: combustion, co-combustion, pyrolysis and gasification.

The biochemical conversion comprises two main types: anaerobic digestion and aerobic digestion.

It uses wet products. They involve microbiological processes that have the effect of degrading the plant material:

- the methane fermentation produces biogas, a mixture of carbon dioxide (30–35%) and methane (50–65%), good quality fuel. The reactions take place at temperatures between 20 and 70 °C;
- the alcoholic fermentation transforms carbohydrates into ethanol;
- the acetono-butyl fermentation produces, under the action of certain bacteria, a mixture of butanol, acetone and ethanol.

The thermochemical conversion is divided into combustion and co-combustion (air in excess), gasification (lack of air) and pyrolysis (in the absence of air). Thermochemical processes are suitable for dry materials such as wood and straw.

Combustion (T ~ 1900 °C) is the oldest and probably most used conversion mode for both domestic and industrial uses. Its efficiency is good insofar as the fuel is rich in structured carbohydrates (cellulose and lignin), and especially dry enough (humidity below 35%).

The co-combustion consists in simultaneously burning a fossil fuel, usually coal, and some biomass (up to 15%) to reduce, in an existing boiler, the amount of initial fuel.

The C/N ratio is the ratio of the amounts of carbon and nitrogen in biomass. It varies from 10 to about 100. Pyrolysis allows to convert relatively dry biomass (humidity below 10%) of C/N ratio greater than 30, in various high LHV fuels, storable, gaseous, liquid and solid (charcoal). It takes place at temperatures between 400 and 800 °C, and can be done in several modes: slow pyrolysis or char (solid product only), conventional pyrolysis at moderate temperature (600 °C) (1/3 gas, 1/3 liquid, 1/3 solid), fast pyrolysis at 500–600 °C, which takes place in about one second (70–80% of liquid), flash pyrolysis at over 700 °C and in less than one second (more than 80% of liquid).

The gasification of biomass (T ~ 800 °C–1000 °C) is obtained by performing a combustion with lack of air in schematically two main stages: pyrolysis producing gas, liquid and solid phases, followed by the gasification itself for the last two phases. It produces a gas called "poor" because of its low heating value (1 kWh/m^3 against 10 kWh/m^3 for methane). By replacing air with oxygen, a synthesis gas ($CO + H_2$) is obtained and used for the manufacture of methanol.

In a gasifier, the fuel is first dried and then pyrolyzed, these two steps being endothermic. The gaseous products are then burned at high temperature, releasing heat, a part of which is used by the previous two steps. The exhaust gases are then put into contact with the solid phase after the pyrolysis and with water from the drying, which causes a reduction reaction that leads to the formation of a synthesis gas rich in CO and H_2, whose LHV is approximately 70 to 75% of that of the original biomass.

Several gasification processes exist:

- fixed bed systems, which consist of two main technologies: the co-current (downdraft) and the counter-current (updraft) gasifiers;
- fluidized bed systems, which consist of three categories depending on the fluidization speed: dense, circulating and entrained fluidized beds.

The first match small or medium facilities, and the latter large ones.

2.4.6.4 *Wind power*

Wind and hydropower are indirect solar energy generated by the movement of air masses and water resulting from the cyclical warming of the Earth by the sun.

Wind energy is the kinetic energy of the air mass moving between areas at different pressures. Wind energy has been harnessed since ancient times, either in horizontal or vertical axis wind turbines, or to propel sailing ships.

The windmill uses are many, with, in order of increasing installed capacity:

- maritime signaling (30 W to 5 kW);
- telecommunications (60 W to 5 kW);
- battery charging in developing countries;
- pumping and desalination of water (1–5 kW);
- domestic use (several kW);
- electricity generation in wind farms of increasing power (up to approximately 200 MW) connected to the network and debiting when it is windy. More than 20,000 machines are installed in the world, with a capacity close to 100 GW, of which more than 25% in the United States and 65% in Europe.

Since 2000, the global installed capacity of wind power has increased from 30 to 40% per year, reaching almost 120 GW in 2008, electricity production being estimated at 260 TWh [IEA Technology Roadmap, 2009].

The most powerful machines have a rated power of a few MW. The cost of land windmills is about $1.5 million/MW offshore and that of wind turbines $2.2 million/MW.

2.4.6.5 *Hydropower*

Hydraulic energy is the potential energy that water gets from altitude. It results from the action of sunlight through the water cycle, evaporated from the surface of the earth, then condensed as rain. It can be large hydro, which will be discussed in the section on electricity, and micro-hydro, or low-power hydro (less than 8 MW).

Small plants, as large hydro plants, can be installed on vertical waterfalls of highly variable height of fall (1.5 to 400 m), and significantly contribute to local electricity or mechanical power needs. Historically, water mills have played a major role in industrial and commercial development.

2.4.6.6 Energy from the sea

Energy of the sea comes from different sources:

- the sun, either indirectly due to the wind that gives rise to waves, or directly due to the thermal gradient between the surface of warm seas and deep water, which is used to operate a low temperature difference thermodynamic cycle. This is known as Ocean Thermal Energy Conversion (OTEC). Conversion devices are still, in both cases, at the prototype stage;
- gravitation, because of the relative movements of the Earth moon and sun, creating tides. This is called tidal energy. However, few sites can cross the threshold of profitability for these facilities. In France, the La Rance plant was commissioned in 1966. It has 24 bulbs and produces 10 MW 0.5 TWh/year (Figure 2.17).

We will only develop here the ocean thermal energy conversion. OTEC cycles are designed to generate electricity in warm tropical waters using the temperature difference between water at the surface (26–28 °C) and at depth (4–6 °C), from 1000 m. (Figure 2.15).

Two main types of cycles are used: closed cycles and open cycles, invented by two Frenchmen, respectively Jacques d'Arsonval in 1881 and his student Georges Claude in 1940, who proceeded to a first experiment.

In all cases, the need to convey very high water flow rates and pump cold water at great depth induces significant auxiliary consumption. Optimization of an OTEC cycle is imperative to take into account those values.

Although technically valid, OTEC cycles are not yet economically viable. Prototypes of various capacities have been realized or are being considered, including in Hawaii and Tahiti.

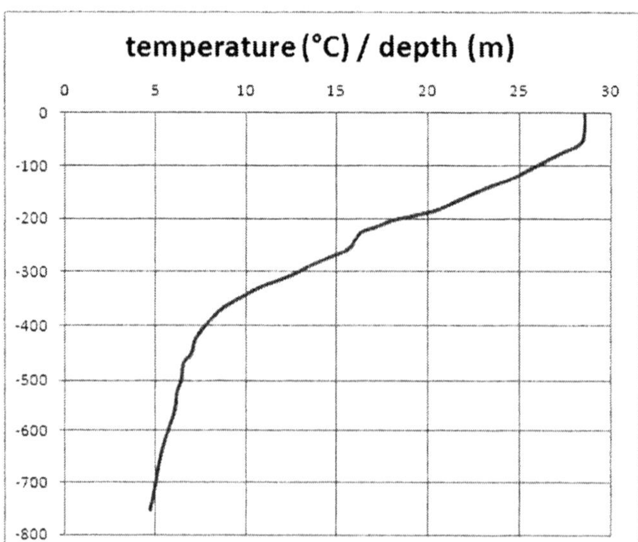

Figure 2.15 Ocean thermal gradient.

Closed cycles use hot water at about 27 °C to evaporate a liquid that boils at a very low temperature, such as ammonia or an organic fluid. The produced vapor drives a turbine, and is then condensed by heat exchange with cold water, at about 4 °C, from deeper layers of the ocean.

The thermodynamic cycle is an intermediate Hirn cycle, whose study poses no particular problem. The sizing of heat exchangers is of course even more critical given the very small temperature difference between hot and cold sources. The values of pinches should be as low as possible while remaining realistic.

In open cycles, the warm water at about 26 °C is expanded in a low-pressure chamber (called flash), which allows to evaporate a small fraction (around 5%). The produced steam drives a turbine and is condensed in a low pressure chamber by heat exchange with cold water at about 4 °C from deeper layers of the ocean. The condensate is virtually pure water, which can be used for food (Figure 2.16).

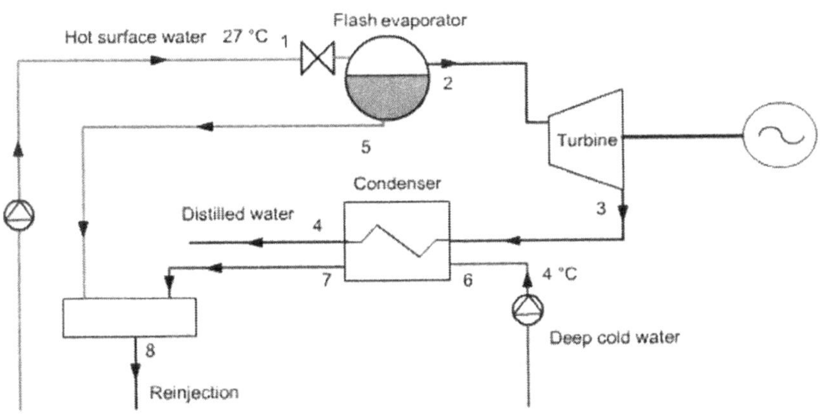

Figure 2.16 OTEC single flash open cycle.

Figure 2.17 Turbine of the La Rance tidal power plant. [*Source*: EDF].

The open cycle thus has the advantage of producing both electricity and fresh water, but the very low expansion ratio involves implementing very large turbines.

Figure 2.16 shows the cycle operation. It involves five elements: a flash evaporator, a turbine, a condenser, a basin for collecting the used seawater, and a vacuum pump.

Hot water is pumped at the surface and brought to a certain height (1), then injected into the evaporator where there is a slight depression determined by the height of the water column between the evaporator and the collection basin. Because of the pressure difference, the water undergoes an isenthalpic throttling (flash) and a small fraction is vaporized (2), then led to the turbine (2–3).

In the condenser, pressure is lower than in the evaporator, thanks to the vacuum pump and the height of the water column between the condenser and the collection tank. The turbine expands the steam produced in the evaporator, producing mechanical power. The vapor is then condensed (3–4) by exchange with cold water, producing fresh water.

The hot and cold water mixed in the collection basin (8) are fed back into the sea at a depth of sixty meters.

2.4.6.7 *Energy of draft animals*

In developing countries, food is generally produced, processed and transported through energy of men and animals. Thus energy of draft animals is and will be for a long time the main source of energy to ensure farm work and transportation in these countries. The conversion efficiency could be greatly improved if farming, food and health equipment were modernized, and if they were incorporated in other rural systems such as the production of grain, meat, milk, fiber, fuel and fertilizer (animal waste).

2.5 ELECTRICITY PRODUCTION

Given its many benefits, electricity continues to experience a strong growth in the world (annual growth of 7.5% from 1963 to 1973, 4.3% in 1975–1986 and 3.1% from 1990 to 2007), more than twice the rate of the growth of primary energy consumption since the oil shocks (see Chapter 6).

Electricity can be produced in many ways, from:

- chemical energy (batteries, fuel cells);
- coal or uranium (steam power plant);
- gas or oil (steam power plants, gas turbines, combined cycle, diesel);
- renewable (hydro, wind, photovoltaic, solar thermodynamic) energy.

Batteries, solar cells and wind turbines can produce electrical energy locally, but they now represent a very small fraction of installed capacity, which is derived essentially from power plants.

Power plants convert into electricity either potential energy (hydropower) or heat (thermal and nuclear power plants). In almost all cases, the plant comprises a generator driven by a turbine.

The hydro turbines are of various types depending on the height of fall: impulse Pelton turbines (Figure 2.18) for heights between 200 and 1500 m, reaction Francis turbines (between 30 and 200 m, Figure 2.19) or Kaplan turbines for low hydraulic

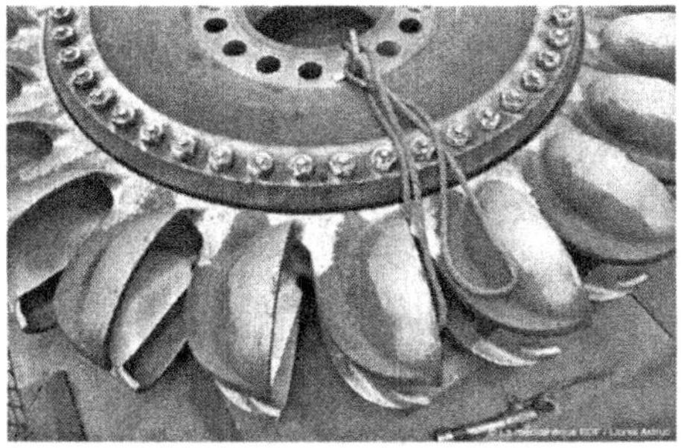

Figure 2.18 Pelton wheel. [*Source*: EDF].

Figure 2.19 Francis turbine wheel. [*Source*: EDF].

heads[4] (below 30 m). Efficiencies of hydropower plants, which convert mechanical energy into electricity can be higher than 85%.

2.5.1 Thermal power plants

Thermal power plants can be of several types.

Steam plants are the most common. They implement a thermodynamic fluid, most commonly water, and contain, besides the alternator, four components:

4 The « hydraulic head » is the technical wording to describe the height of the column of water upstream a hydraulic turbine or downstream a water pump.

- the boiler, hot source of the cycle, which, depending on the case, may be a combustion boiler (flame plant), or a nuclear boiler;
- the turbine, comprising a large number of stages, grouped by high, medium and low pressure bodies, wherein the steam expands providing mechanical power;
- the condenser, heat sink of the cycle, which condenses the working fluid being cooled, either by moist air in a cooling tower or by water from a river or the sea;
- the pump, which, at the cost of negligible mechanical work, pressurizes water leaving the condenser.

Flame steam plants operate at high pressure and temperature (respectively of the order of 160 bar and 560 °C), much higher than those of nuclear power plants for security issues (respectively about 55 bar and 270 °C). Units reach a capacity of 600 MW for flame power plants and up to 900 MW and even 1450 MW for nuclear power plants. Their efficiencies are between 30 and 40%.

Gas Turbines (GT) or Combustion Turbines (CT) facilities are most often used in open cycle (Figure 2.20), and sometimes in closed cycle. They consist of three basic components:

- a compressor, usually centrifugal or axial, which is used to compress the ambient air at a pressure of between 10 and 30 bar in modern machines;
- a combustion chamber in which fuel is injected under pressure, combusted with air previously compressed (the latter in large excess in order to limit the exhaust gases temperature at the turbine inlet);
- a turbine, usually axial, in which the high temperature gases exiting the combustion chamber are expanded. A significant proportion (60–70%) of the work generated by the turbine is used to drive the compressor.

To achieve compression ratios *r* of 20 or 30, the compressor is multistage, with at times intermediate cooling to reduce the work consumed. Axial rotors are made of a stack of discs, either mounted on a central shaft, or drum assembled on their periphery. The materials used are aluminum or titanium alloys for the first stages and steel alloys and refractory alloys for last stages that can withstand temperatures up to 500 °C.

The combustion chamber is normally made of a refractory alloy.

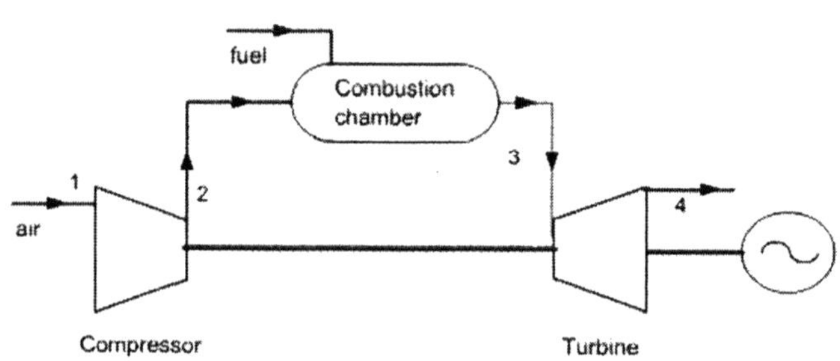

Figure 2.20 Schematic of a gas turbine.

In a gas turbine open cycle, the main technological constraints are at the first stages of the expansion turbine, through which the exhaust gas flows at high temperature.

The parts most at risk are especially the rotor blades, which are very difficult to cool and in addition are particularly sensitive to abrasion. It is therefore important to use a very clean fuel (no particles and chemicals that may form acids), and limit the temperature depending on the mechanical characteristics of the blades.

The materials used for turbine blades are refractory alloys based on nickel or cobalt, and manufacturers intend to make use of ceramics in the future.

Two broad categories of gas turbines are generally distinguished: industrial gas turbines, heavy and robust, but of average performance (efficiency η between 28% and 38%), and gas turbines derived from aviation or "aeroderivative" which are much more efficient and lighter (η between 35% and 42%), but also more expensive. The capacities of the first range from tens of kW (microturbines) to several hundreds of MW, while those of aeroderivative machines are generally between a few hundred kW to tens of MW, corresponding to those of aircraft engines. Most jet engines used in aviation today are in fact variations of the gas turbine open cycle. The aviation market has helped fund major technological development programs on these engines, which led to the development of highly efficient gas turbines, which could supplant industrial gas turbines, or allow turbine design consisting of enhanced efficiencies and low cost, including components of existing jet engines for high pressure compressor and turbine sections, and industrial parts for low-pressure sections.

Recent improvements have allowed GT efficiencies to go up to 33% for large power units (200 MW), and they are expected to reach nearly 40%.

One way to enhance the residual enthalpy of the exhaust gas is to use them as a heat source for a second cycle of production of mechanical energy. This new generation of power plants is called Combined Cycles (CC) (Figure 2.21). The excellent efficiencies reached today by combined cycle power plants (above 60% LHV), are the result of integration into a single production unit of two complementary technologies in terms of temperature levels: gas turbines, which operate at high temperature (in an aero-derivative turbine gases typically enter at 1300 °C in the expansion turbine, and come out at around 500 °C), and steam plants, which operate at lower temperatures (between 450 °C and 30 °C in this case).

Gas turbine combined cycles have the added benefit of being a very modular, fast-building and inexpensive investment (€300/kW for the GT, €500/kW for the CC,

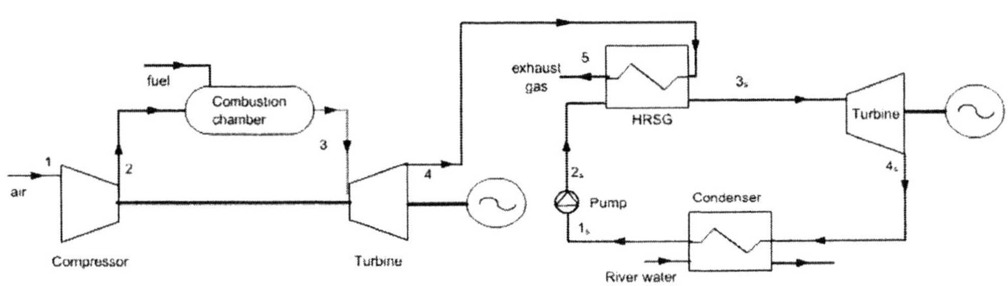

Figure 2.21 Schematic of a combined cycle.

compared to €1,000/kW for a coal-fired fluidized bed or over €1,400/kW for a nuclear power plant).

Finally, generators powered by diesel engines are used to generate electricity when the installed capacity is low (less than a few MW), and therefore off-grid especially in remote areas. Their maintenance is easy and the installation prices relatively low, but they have the disadvantage of having high operating costs.

2.5.2 Production management

One of the major constraints in the production of electricity is undoubtedly the difficulty of storing the produced energy. Electricians must be able to adjust their production to meet the demand, which fluctuates strongly, either daily or on a seasonal basis. For example, Figure 2.22 shows the values of peak demand on the EDF network in France on January 19, 1989 and throughout the year 1989.

It is essential to ensure the continued adequacy of production to meet demand, otherwise complete network outages may occur. The most famous ones are those of August 2003 in the United States and Canada (which resulted in the shutdown of a hundred power plants in the Great Lakes region and the East coast), of November 2009 in Brazil (which affected a quarter of the country's population), and of July 2012 in India (a giant electricity blackout affected half of the country, after the almost simultaneous collapse of systems serving twenty states in the north, east and northeast).

The load curve therefore shows significant fluctuations, which can be synthesized on the monotonic load curve, as described in section 5.4.2.

Two basic strategies can be used to meet this requirement:

- trying to reduce the load curve amplitude of variations by offering rates for users encouraging them to use off-peak electricity and divert peak power demand;
- better managing the fleet of plants to minimize operating costs.

Firstly, the natural storage offered by lake plants is used by preferably operating these plants at peak times. Plants whose cost of fuel is lower are used during periods of base demand, reserving for the peak those whose investment cost is lower, even if the fuel cost associated with them is very high. Finally, energy that has

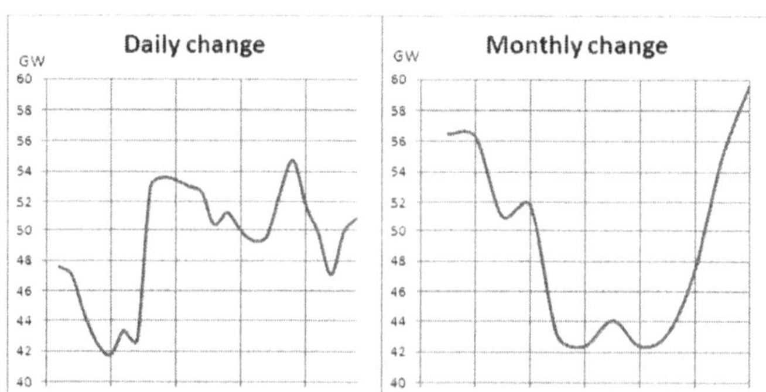

Figure 2.22 Load curves.

a very low base cost (e.g. nuclear) is used to store water in lakes during low demand hours, and turbine it during peak hours (with a cycle efficiency of about 70%).

2.5.3 Transport and distribution

As electricity cannot be easily stored, it must be distributed instantly. This is the role of the network, which is composed of four sub-networks based on the current voltage, connected together by switching stations. Joule losses can be up to 7% of the total production.

Placed at the outlet of the plant, lines at Very High Voltage (VHV) transport electricity set at very high voltage (400,000 and 225,000 volts) over long distances. All VHV lines form an interconnected network allowing for the discontinuation of certain plants.

Production and energy movements are managed by dispatching control centers. These centers, assisted by powerful computers, establish production schedules based on forecasts and real-time data corrected to deal with contingencies.

The VHV network is connected to the networks of neighboring countries, which has the effect of increasing supply security in the event of a catastrophic failure, and allows for international trade in electricity.

High voltage lines (90,000 and 63,000 V) are, with VHV lines, the transport network. They are used to route electricity to distribution utility centers and large users (railway, chemical industries, steel, metallurgy).

The medium voltage network (between 15,000 and 20,000 V) supplies medium size customers from distribution centers. In addition, it serves low-voltage customers through MV/HV transformers.

Low voltage lines (380 or 220 V) form, together with the MV lines, the distribution network. LV network supplies millions of domestic users.

Each of the four previous networks is connected to its neighbors by transforming stations, home breakers, designed to protect against network malfunctions, isolators, which allow, if necessary, to disconnect part of lines and transformers, which modify the voltage junctions between different networks.

2.5.4 Cost structure

Investment costs (expressed in 2010 U.S. dollars) range from $730/kW for a gas turbine, to $2,130/kW for coal and $4,000/kW for nuclear power plants, as shown in Table 2.4 [adapted from Hansen & Percebois, 2010].

Table 2.4 Production costs of electricity.

	Unit size MW	*Cost of invest. $/kW*	*Fixed costs $/kW*	*Operating costs $/MWh*
Gas turbine	150	730	64	140
Combined cycle	480	1,070	95	84
Pulverized coal	750	2,130	180	42
Nuclear	1,400	4,100	330	14

The production cost consists of fixed costs related to depreciation of equipment, operating costs, considered to be independent from the amount of produced electricity and operation costs, highly variable: \$140/MWh for gas turbines, \$42/MWh for coal, and \$14/MWh for nuclear. The cost of production must be increased by costs of transmission and distribution, which more or less doubles the total price.

2.5.5 Conclusion

Electricity has specific uses, such as lighting, powering electric motors, electrolysis, induction heating or acquisition, processing and transmission of information. It must be distributed to the places of consumption, even if the production is often centralized since it is very sensitive to scale effects. This requires the development of a complex and expensive network. Besides, the almost non-storable nature of electricity leads the producer to continuously monitor the evolution of demand, which requires that the fleet be sized to the maximum power demand, and thus induces significant additional costs. All these constraints are combined to generally give the power industry a centralized and rigid character.

But there is also room for decentralized production that can help secure the interconnected network, even if they have special operating constraints including frequency stability.

2.6 THERMAL OR PNEUMATIC ENERGY STORAGE

Energy is generally very difficult to store, except in the form of fuels (especially liquid ones). In particular, it is still impossible to economically store large amounts of electricity, so that power plant park managers have to adjust their production to meet a demand that fluctuates greatly, both daily and depending on the season. However, this way of doing things has its limits and induces a cost, so that electricity prices vary widely depending on the time (Figure 2.23).

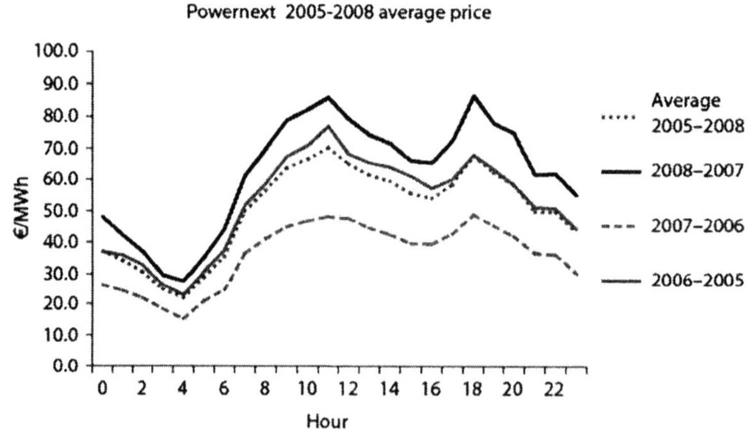

Figure 2.23 Changes in the average price of electricity during the day.

In this section, we will only focus on the problem of large scale thermal or pneumatic energy storage without addressing "small" storage units such as batteries, flywheels, etc.

In addition to the hydraulic storage system mentioned in section 2.5.2, several systems can be implemented to foster off-peak consumption by producing electricity during peak hours, for instance by compressing some air when electricity is cheap and expanding it in a turbine when it is expensive The association of such a compressed air storage with a thermal storage can improve the system performance.

Thermal energy storage can take several forms:

- sensible heat: water, synthetic oils, pressurized steam, molten salts without phase change, ceramics, concrete, etc.;
- latent heat: solid-liquid, liquid-vapor, solid-vapor or solid-solid phase transition for a variety of materials (water, paraffin, salts, metals, etc.);
- in thermochemical form through endo/exothermic chemical reactions, adsorption, absorption.

Pneumatic storage consists in compressing gas in the storage phase, then expanding it during the retrieval phase. The gas most often used for this is air.

Both forms of storage can be combined to achieve a better performance.

Thermal or air storage effectiveness is always relatively low, so that the use of such storage is mainly justified in two cases:

- when the availability of the resource is variable and not synchronized with the needs. This is typically the case for solar installations;
- when the price of energy varies greatly from one period to another: it may then be worthwhile to store energy during off-peak hours and destock it during peak hours.

In any case, during the retrieval phase, thermal or pneumatic storage systems do not restore energy in exactly the same state (in the thermodynamic sense of the word) as during the storage phase. Losses are reflected by lower temperature and pressure; it is important to know that to determine the functioning of the entire system. However, transfer functions between the stored energy and energy removed from storage are often poorly known.

After a brief methodological presentation of the issues at stake, this section will focus on storage systems that are either already widely used, especially for cooling, or still under consideration.

The introduction of storage in a thermodynamic system poses a number of methodological difficulties. Quite intuitively, storage-release operations induce losses that can rarely be assessed by defining a simple effectiveness equal to the ratio of the released energy to the stored energy, as the energy quality usually drops, and also depends on how the storage is carried out.

Let us consider the simple storage by sensible heat of a liquid thermal fluid (water, oil etc.). The modeling of such a storage is usually made by considering two limiting cases, called "perfectly mixed storage" and "plug flow". In the first case, we assume that at any time the temperature is uniform throughout the tank, set at the average value and depending on the temperature before injection of fluid and on the amount of fluid added. In the latter case, it is assumed that there is a perfect layering in the tank between the fluid that was initially present and the stored fluid (with the assumption that hot fluid is injected in the upper part of the tank).

Moreover, one can consider heat losses from tank to the outside, usually assumed to be proportional to its surface and the temperature difference between the stored fluid and outside temperature.

It is clear that these two cases correspond to extreme situations, the first leading to a much larger decrease of the energy quality than the second. The actual situation is certainly somewhere between the two: some mixing exists, but can be minimized by taking technically-sound measures. It is thus possible, when you have a large storage tank, to envision it as modules connected to each other. Each module can then be considered mixed, while a very good stratification exists between different modules at different temperatures.

In pneumatic storage, a compressor fills a storage chamber by injecting a gas whose pressure and temperature vary depending on the storage state, because only the assumption of a perfectly mixed medium may be used. The compressor performance and thus the energy cost of the operation also directly depend on technological characteristics and storage conditions. In release mode, pressure and temperature drop, also according to complex laws.

The main types of existing or planned thermal or pneumatic energy storage are:

- storage of cold nodules in phase change, such as latent heat storage systems called STL, marketed by Cristopia (www.cristopia.com). A STL consists of a tank filled with a heat transfer fluid and nodules containing a phase change material whose melting point can be chosen between –33 °C and 27 °C. Approximately 60% of the volume of the tank is occupied by the nodules and the remaining 40% by the heat transfer fluid (ethylene glycol or brine). Since industrial refrigeration systems must produce cooling energy often with very important loads for short durations (corresponding to production cycles), STLs have many advantages such as significant reduction of cooling capacity installed (up to 70%), use of electrical energy at the lowest rates for savings, and management of refrigeration according to actual needs;
- Saipem Company has patented a storage method called THESE (Thermal Energy Storage of Electricity), which implements heat regenerators filled with refractory materials that store energy, and heat engines (turbines and compressors) used for energy conversion. The storage method is based on innovative technology, which combines a heat pump during the storage phase and heat conversion thanks to gas turbine cycles. During the storage phase electricity is used to drive a heat pump that transfers heat from one chamber to another at a higher temperature, and during the release phase, heat is turned back into mechanical energy by a heat engine;
- the best known energy storage system is called CAES (Compressed Air Energy Storage), or its variant AACAES (Advanced Adiabatic CAES). It requires large underground caverns, which is possible only in places where the geology is favorable (deep salt domes for example, or highly permeable deep aquifers) and therefore rarely located in appropriate areas. Moreover, except for AACAES, this technology requires combustion of natural gas during release. In such a device, air is stored under pressure during peak hours in a large cave. In production

mode, this air is used as an oxidizer in a gas turbine, substantially reducing the compression work, which represents 60–70% of the work generated by the turbine in classic GT cycle (see above). The available capacity can be tripled compared to operation without storage;

- the use of isothermal processes allows for good efficiencies. Using an isothermal cycle at room temperature has the advantage of not requiring heat storage, eliminating losses and allowing for indefinite storage. This is the basic idea that has been studied at the École Polytechnique Fédérale de Lausanne (EPFL) and led to the concept of "Oleo-Pneumatic Battery" or OPB.

2.7 COMMON CHARACTERISTICS AND COMPARISONS

The previous sections presented the main energy sectors (oil, natural gas, coal, synfuels, nuclear power, renewable energy, thermal or pneumatic storage), i.e. how the main sources of energy can be exploited.

They have shown the existence of numerous competing technologies that can be implemented in accordance with specific situations. Choosing between them depends on many factors which will be considered in the following chapters, and is a focus of energy policy, as will be discussed in Chapter 5.

Finally, we have deliberately devoted little attention to two recent developments hitherto, as they will be addressed in more detail in Chapter 8: energy's impact on the environment and technological changes.

In this final section, before comparing the various types of energy, mainly from an economic viewpoint, we will mention some characteristics generally shared by these sectors:

- their delays in their implementation;
- their diversity and rigidity;
- their social acceptance.

2.7.1 Implementation time

The graph in Figure 2.24 compares the time of implementation of various types of energy. It clearly shows that these durations are always very long (5–15 years).

2.7.2 Diversity and rigidity

From all this, two salient points emerge: the diversity of energy systems and their rigidity:

- a multiplicity of energy technologies are indeed available to satisfy a given energy need, leading to very different national energy mixes. Some rely on a highly centralized industry, related to the high level of investment that is required (exploration and production of oil and gas, coal, nuclear, electricity sector, geothermal, etc.), while others can be widely distributed (oil distribution, solar energy, wind, small hydro, etc.). Depending on the available resources, on the nature and volume of the needs, on the degree of technological

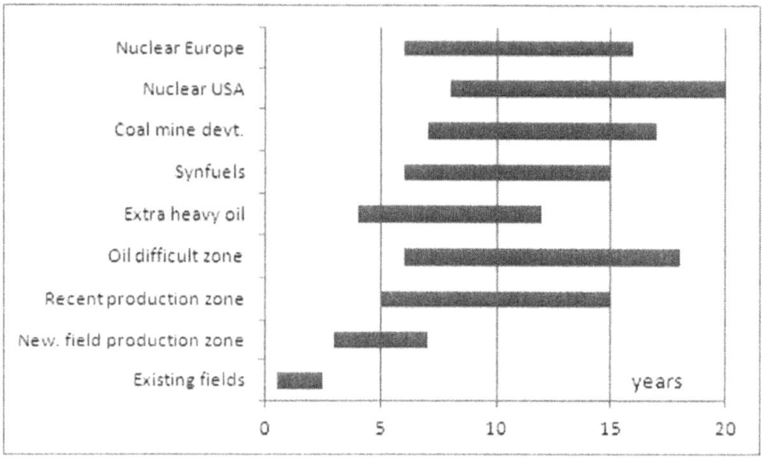

Figure 2.24 Time of implementation.

development of the country and on its environmental constraints, it is obviously possible to use all or part of these technologies, so that the solutions may vary widely from one country to another. This is the purpose of energy planning (see Chapter 5), which aims at determining the best energy mix based on the existing context;

- the diversity of technologies offers a broad range from which to choose during the development of energy policies, but once these choices are made they affect the future in the medium and long terms, which introduces strong rigidities.

These rigidities are related to several factors:

- the complexity of energy systems;
- the concentration of commercial energy productive sectors;
- the high investment costs of generation, transmission and distribution;
- the long delays in the implementation of energy projects.

This rigidity is not new and explains why the substitution time between different energy sources is still historically quantified in terms of decades (5–10), as shown in Figure 2.25, which presents a historical overview of the evolution of the share of each of the main sources of energy in the global mix.

It can be seen from this graph that after long periods of progressive substitutions, the share of the different energy sources appears to have stabilized over the past forty years.

For the future, it is hardly conceivable that the cycles highlighted above will continue so regularly. We will see later that it is particularly likely that coal will experience a second wind, given its sizeable reserves.

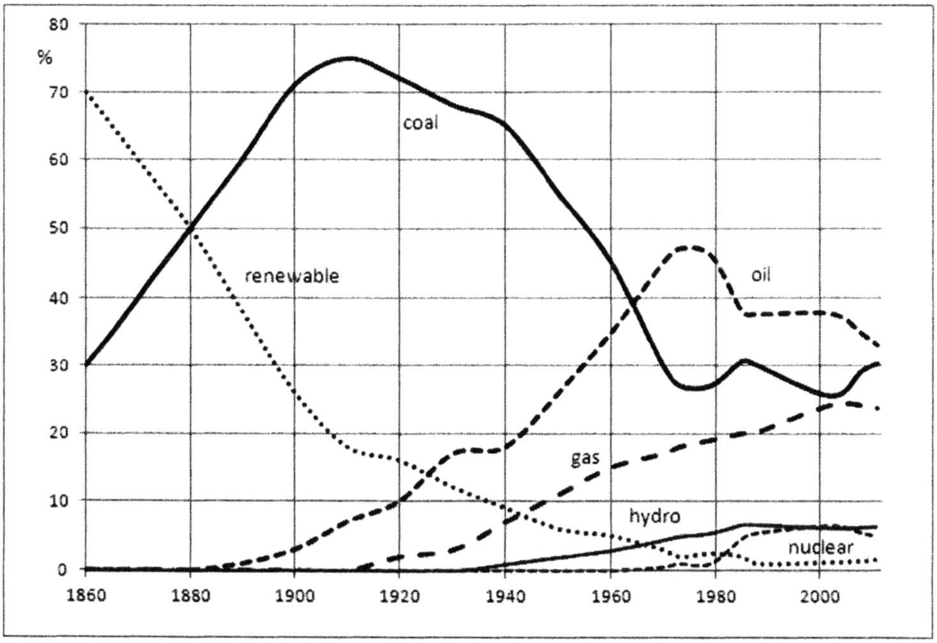

Figure 2.25 A retrospective of shares of the main energy sources.

2.7.3 Social acceptance

The social acceptance of energy technologies, which did not pose particular problems for many decades, is being increasingly questioned, as it is the case for many industrial projects.

In a growing number of countries, civil society is concerned about the possible consequences of energy projects, particularly in terms of systematic and accidental pollution.

This is the so-called "Not In My BackYard" or NIMBY syndrome, when people object to potentially harmful projects being installed close to their home. Such objections have arisen for many technologies, whether nuclear power plants, wind farms, dams, power lines at high voltage (VHV), LNG terminals, or more recently shale gas production facilities.

This phenomenon, which increases implementation time and the rigidity of chains, fits into the logic of the precautionary principle and can lead to the "BANANA" (Build Absolutely Nothing Anywhere Near Anything) attitude, and to abandoning any development plan for political and technical reasons.

The development of energy infrastructure should, given the necessary delay, be decided long before shortages occur and call for immediate implementation, and the fears and reluctance of civil society must be taken into account early enough, by setting up appropriate consultation mechanisms. This is one of the main energy policy axes, presented in section 5.6.1.

2.7.4 Comparisons between sectors

2.7.4.1 *Economic comparisons*

In economic terms, it is possible, under certain assumptions, to compare different kinds of energy. A first difficulty arises, however, because some of them (e.g. oil) entail low investment costs and high operating costs, while the reverse is true for others, including renewable energies. In order to compare them, it is therefore necessary to conduct a cost–benefit analysis on the duration of the project and hence to make assumptions about the equipment life and the appropriate discount rate. To refine things, it is also desirable to take into account the external costs or costs of externalities, i.e. the costs caused by the industry but supported in whole or in part by one or several other economic agents. These costs include in particular those related to public health, impaired by pollutants and the effects of global warming.

For conventional energy, a number of factors regarding the technical costs were given in the preceding pages.

For renewable energy, these costs are subject to rapid changes in technology and in market size.

Figure 2.26 provides the cost of electricity from renewable sources, calculated by amortizing over 15 years investment costs [Resh et al., 2006]. Photovoltaic energy does not appear because it is outside the considered scale (cost between $570 and $2,200/MWh. Values provided by the authors were converted from euros to dollars by adopting an average conversion rate of €0.75/$.

The graph in Figure 2.27 compares the production costs of different hydrocarbons, including non-conventional fuels [IEA World Energy Outlook, 2008].

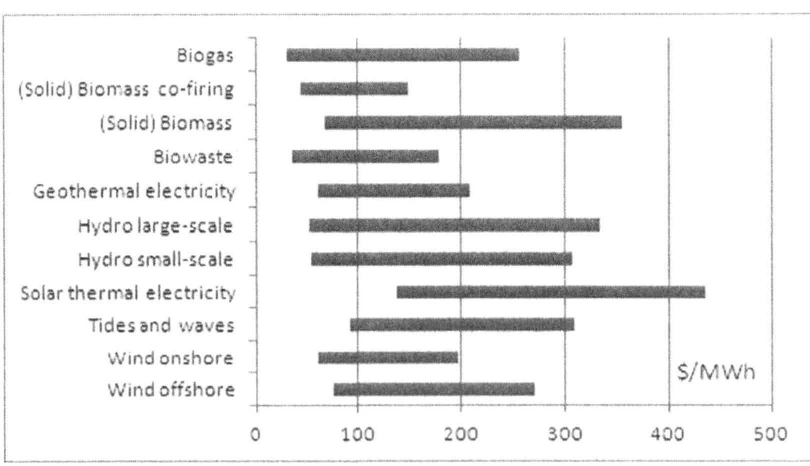

Figure 2.26 Costs of electricity from renewable sources.

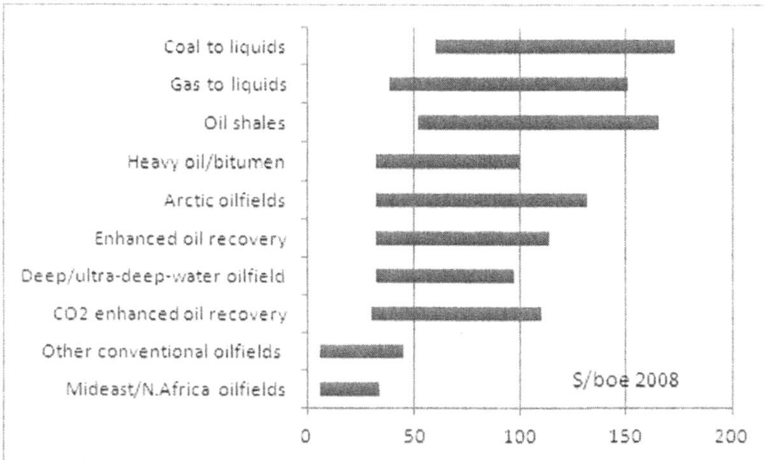

Figure 2.27 Comparative costs of production of hydrocarbons.

The advantages of oil are evident in these graphs. They also show that at a price of $30/b, a number of competitors are beginning to appear on the market.

2.7.4.2 *Other comparisons*

Economic evaluations are not enough to make comprehensive comparisons between different energy systems.

To make the best possible choice between them, new methodological tools have been proposed, in particular to quantify their overall impacts. Among these, let us mention Life Cycle Analyses (LCA), "from well to wheel" emissions analyses, and carbon footprints.

A life cycle analysis allows one to estimate the total energy and amount of raw material that are spent from the design phase of a product to its recycling or destruction, i.e. from the extraction of these raw materials to their disposal at the end of the product's life, including the phases of distribution and use. An example of the type of calculation provided by this approach is given by equation (8.1) in Section 8.2.1.1, which assesses the impact of complex technology such as that used for refrigeration on the greenhouse effect.

We will not develop these complex methodological tools further in the context of this book.

2.8 CONCLUSION

The mind map of Figure 2.28 gathers the main concepts pertaining to energy sectors.

Figure 2.28 Mind map of energy sectors.

REFERENCES

Babusiaux D., Ed, Recherche et production du pétrole et du gaz: réserves, coûts, contrats Technip, 2002, ISBN 2-7108-0791-2.
Gicquel R. Energy Systems: A New Approach to Engineering Thermodynamics, January 2012, CRC Press, ISBN-13: 978-0415685009.
Gomez-Exposito A., Conejo A. J., Canizares Cl., Eds., Electric Energy Systems Analysis and Operation, CRC Press 2008, Print ISBN: 978-0-8493-7365-7, eBook ISBN: 978-1-4200-0727-5.
Hansen J. P., Percebois J., Energie, économie et politiques, De Boeck, Bruxelles, 2010, ISBN 978-2-8041-6177-4.
IEA, Technology Roadmap, Bioenergy for Heat and Power, Paris, 2012.
IEA, Technology Roadmap, Wind energy, Paris, 2009.
IEA, Technology Roadmap, Solar photovoltaic energy, Paris, 2010.
IEA, Technology Roadmap, Geothermal Heat and Power, Paris, 2011.
IEA World Energy Outlook 2008, IEA, Paris, 2008.
IEA World Energy Outlook 2008, IEA, Paris, 2008.
IFP Energies Nouvelles, Consommation mondiale d'énergie dans le secteur des transports routiers en 2009 [Online] Available from: http://www.ifpenergiesnouvelles.fr/media/images/espace-decouverte/les-cles-pour-comprendre/automobile-et-carburants/carburants-conventionnels/consommation-mondiale-d-energie-dans-le-secteur-des-transports-routiers-en-2009/ [Accessed 01/18/2013].
Resch G., Faber Th., Haas R., Ragwitz M., Held A., Konstantinaviciute I., Potentials and cost for renewable electricity in Europe – IEE project OPTRES, Vienna, February 2006.
Saniere A., Serbutoviez S., Silva C., Les investissements en exploration-production et raffinage, IFP Energies Nouvelles, Rueil Malmaison, 2010.
UNCTAD, Review of Maritime Transport 2011, New York and Geneva, 2012.
WNA, The Nuclear Fuel Cycle: Education, World Nuclear Association, [Online] Available from: http://www.world-nuclear.org/education/nfc.htm/[Accessed 09/18/2012].

FURTHER READING

Ballerini D., Alazard-Toux N., Les biocarburants, État des lieux, perspectives et enjeux du développement, Technip, Paris, 2006, ISBN 2-7108-0869-2.
Boy de la Tour X., Technologies pétrolières: les nouvelles frontières, Revue de l'Énergie, Paris, n° 456, février 1994.
Energy Storage. A key technology for decentralised power, power quality and clean transport, European Commission, Report EUR 19978, 2001, ISBN 92-894-1561-4.
Guide de l'Energie, La Documentation Française, 1988.
Hafner M., Gaz Naturel et Électricité, Analyse technologique et économique de la génération d'électricité et du transport de gaz pour les pays du bassin méditerranéen, Thèse de Doctorat, École des Mines de Paris, 1994.
Hafner M., Karbuz S., Esnault B., Elandaloussi H., Long-term natural gas supply to Europe: Import potential, infrastructure needs and investment promotion, 20th World Energy Congress, 11–15 November 2007, Rome, Italy.
Hafner M., Tagliapietra S., Elandaloussi H., Outlook for Electricity and Renewable Energy in Southern and Eastern Mediterranean Countries, Energy, MEDPRO Technical Papers, 2012.
La petite Encyclopédie de l'Énergie, DGEMP, Dunod, Paris, 1988.
Maillet P., Cassette-Carry M., L'Énergie, Que Sais-je, PUF, 1982.
Odru P., Coord, Le stockage de l'énergie, Dunod, Paris 2010, ISBN 978-2-10-054042-6.

Chapter 3

Macroeconomic indicators and accounting of energy

3.1 PRESENTATION

This chapter introduces key indicators that will be used in the rest of the book to study the macroeconomic impact of energy policies: GDP, GNP, primary energy consumption, final energy intensities, etc. Its last section presents the main characteristics of the different energy markets.

It should be noted that, given the scope of this book, it focuses on indicators related to energy rather than CO_2 or greenhouse gas emissions. As explained in section 3.7, these two types of indicators are highly correlated for most countries, so that the conclusions presented for one are generally valid for the other.

As demographics obviously play an important role in the evolution of these quantities, let us first examine how the world population is likely to evolve.

3.2 EVOLUTION OF THE WORLD POPULATION

The graph in Figure 3.1 shows the past and projected world population, assuming an average scenario of declining fertility. It should continue to grow beyond 2050 and could within 200 years reach a limit (called the hypothetical stationary population) estimated to be about 11 or 12 billion people, that is to say almost two and a half times the population of 1990.

As shown in Table 3.1, the regional distribution should not vary much, the only significant changes concerning Africa, whose share could increase from 15 to 21.8% from 2010 to 2050, and Europe, which would decrease from 10.6 to 7.6%.

3.3 BASICS

Economics is the science whose object is the knowledge of phenomena related to the production, distribution and consumption of wealth and property in human societies.

Very commonly, a distinction is drawn between microeconomics, which focuses on the detailed study of the economics of a business and individual "economic agents" considered as autonomous centers of decision, and macroeconomics, which aims at

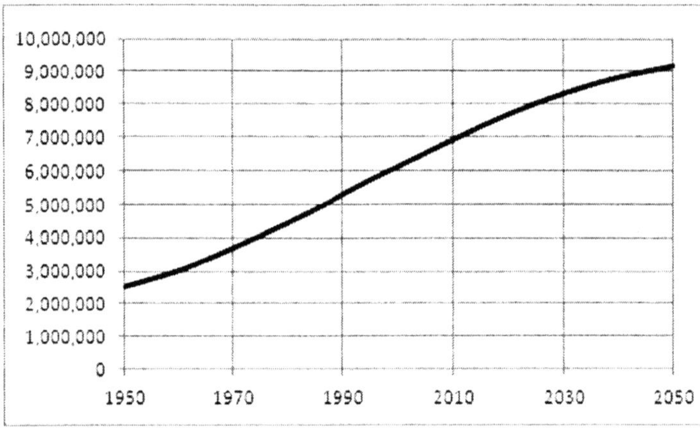

Figure 3.1 World population.

Table 3.1 Share of the main regions in the world population.

	2010	*2030*	*2050*
Asia	60.3	59.2	57.2
Africa	15	18.3	21.8
Europe	10.6	8.7	7.6
Latin America	8.5	8.3	8
North America	5.1	4.9	4.9

understanding the overall result of the behavior of all actors, at the national level for example.

Microeconomics is a basic science to guide the decision of economic agents, who most often seek to reduce their costs and maximize their profits.

Macroeconomics can in turn guide the choice of national decision-makers on policy issues, e.g. monetary policy, taxation, social policy, labor legislation, research or the fight against inflation and unemployment, etc.

Given the significant impact it can have on the wellbeing and social production as a whole, economic policy is now at the center of the political debate.

It has now been recognized for several years that energy must be considered as one of the basic factors of production, along with capital, labor and natural resources.

The study of the relationship between energy and the economy generates a lot of interest, and may sometimes be highly controversial, as in the case of nuclear power.

3.3.1 Gross Domestic Product

To measure a country's level of economic activity, global indicators are needed, which should preferably be comparable from one country to another and from one year to another. Given the multiple forms that economic activity can take, there is still no faithful and unbiased standard measurement.

One of the most commonly used indicators is the Gross Domestic Product (GDP), which is an aggregate indicator measuring the total final output of goods and services of a given national economy. The Gross National Product (GNP) is also commonly used, as it is more representative of the national wealth generated during a year.

The gross domestic product is defined as the sum of the added value of each economy branch, increased by the VAT (Value Added Tax) on products and tariffs.

In this definition, the branches of the economy are all production units producing the same product category, and the added value is the difference between the value of goods or services produced by a company and the goods or services used for their production ("intermediate consumption"). This accounting method allows one to account only once the added value by each participant in the production process of a complex good.

By adopting this convention, it is possible to measure the level of economic activity of a country, but it is not fully satisfactory, for two main reasons:

- GDP measures all the activities giving rise to monetary exchange, be they useful or not: for example, when the number of car accidents or medical costs increase, the GDP grows, etc.
- GDP only measures what is paid and accounted for, so that everything that is not accounted for is ignored (craftsmanship or volunteer work, including at home, "free" use of natural resources, pollution, etc.).

GDP thus has two basic drawbacks. In the case of developing countries with a large informal sector ("moonlighting"), due to the second bias it is clear that GDP can be a very erroneous representation of the actual volume of economic activity. It can therefore be necessary to correct the official values of GDP by various factors to conduct some analyses, as we shall see in section 3.3.3.

3.3.2 Gross National Product, Gross National Income

The gross national product measures the total external and domestic production of a country. It is equal to GDP plus the compensation for labor and capital that residents receive from abroad, and minus the payments made to non-residents in the same way.

GNP values therefore differ slightly from those of GDP. They better reflect the national wealth, rather than the national production. Significant differences can arise for the countries that have significant external financial assets and whose GNP is greater than their GDP, or, conversely, for highly indebted countries, who must draw from their domestic production the amount of loan repayments (capital and interest). For example, Greece's GNP in 2011 was 5.3% lower than its GDP, while in France the two figures were nearly equal (source: the World Bank).

It is important to note that the values of GDP and GNP are representative of the activity during a given year, but not of all the property or wealth accumulated and available in a country. To take these variables into account, it would be necessary to have indicators that do not exist (when a monetary transaction is recorded it is possible to identify the production of a good or service, but there is no record of its life duration nor of how its value changes over time).

This is a reason why, despite rapid economic growth, some countries that have reached high levels of per capita income are still among the developing countries.

Only when their total capital value is comparable to that of industrialized countries will they be considered developed.

Moreover, as the basic purpose of development is to improve the living conditions of populations, it is important to keep in mind that both GDP and GNP very imperfectly take into account social aspects, and that in any case, they should not be used as the sole measure of developmental success. The human development index, which we introduce later, is an attempt to correct this bias.

In conclusion, the available indicators to measure the economic activity of a country are not fully satisfactory, far from it. That being stated, as most often it is impossible to get anything better, one has to try and make the best use of these quantities.

Recently, the GNP was renamed Gross National Income (GNI).

3.3.3 International comparisons

In addition to the uncertainties described above, the apparent simplicity of the figures obtained for GDP (or GNI) masks a number of difficulties: two problems arise when trying to compare GDPs of different countries, or for a country, the GDP growth from one year to the other.

The first is that, in order to compare changes over time in GDP, one must take into account the effects of inflation. For this, one has to correct (or "deflate") it by dividing the GDP by an index whose variation is supposed to represent the evolution of prices. One then obtains values expressed in volume or in constant currency. This is somewhat arbitrary, since it depends on the choice of the GDP deflator.

The second is that GDP is determined from calculations in national currency and then converted into dollars, making use of the average exchange rate for the year. The value obtained thus depends highly on this parameter, which can cause some distortions, especially for the poorest countries.

To compensate for this bias, one may refer to the GDP in Purchasing Power Parity (PPP). PPPs are currency conversion rates that allows one to compare GDPs by eliminating the differences in price levels between countries. The comparison is based on the price of a "basket" of goods and representative services that are comparable across countries. The reference set consists of approximately 3,000 products corresponding to goods and services constituting the GDP.

Moreover, in order to compare the levels of economic activity and wealth in many countries, one must take into account the differences in population. For this, GDP or GNI values are divided by the number of inhabitants in the country, and the result is called the gross domestic product per capita (GDP/capita), or gross national income per capita (GNI/capita), sometimes called per capita income.

The differences that can be observed from one country to another are considerable: from less than 300 euros per year (average per capita income of 25 euros/month) for the poorest countries, such as Eritrea, to more than 30,000 euros per year (2,500 euros/month) for richer countries as the United States. Figure 4.1 shows the values of per capita GDP in a number of countries, including those that are the subject of case studies presented in Chapter 7.

It is customary to form large groups of countries, although this practice is somewhat artificial given the great diversity that exists within these groups. Nowadays, two large blocks are generally considered, sometimes called the "North" and "South". The "North" includes the "industrialized" countries and the "South" all Developing Countries (DCs). We discuss these topics in Chapter 6.

3.3.4 Economic growth and development

Economic growth occurs when a country's GDP increases from one year to another. If it decreases, there is a recession.

For fifty years, economic growth has become one of the main criteria for judging the performance of a country, both internationally and in the national political debate, though its relevance might sometimes be discussed.

Economic growth stems from a continuous excess of global production over consumption and investment. This results in an increase in GDP, which can be measured by its growth rate p:

$$p = \frac{\Delta \text{GDP}}{\text{GDP}} \quad \text{or} \quad \frac{\text{d GDP}}{\text{GDP}}$$

Strictly speaking, these two forms are not equivalent, and, by definition, $p = \frac{\Delta \text{GDP}}{\text{GDP}}$. For clarity of differential calculus, we will express it more readily in the form of $\frac{\text{d GDP}}{\text{GDP}}$, which differs from the previous one only by terms of second order, but has the advantage of being easier to handle in mathematics.

For development to occur, it is not enough that its GDP increases in volume: a necessary (but not sufficient) condition is that the GDP must grow faster than the population, that is to say, the GDP per capita must increase in volume.

If the population growth rate is:

$$d = \frac{\text{d pop}}{\text{pop}}$$

$$\text{then } r = \frac{\text{d (GDP/pop)}}{\text{GDP/pop}} = \frac{\text{d GDP}}{\text{GDP}} - \frac{\text{d (pop)}}{\text{pop}}$$

$$\boxed{r = p - d}$$

The growth rate of GDP per capita is equal to the GDP growth rate minus the population growth rate.

This relationship means that in order to increase the per capita income, production must grow faster than the population. This is a major problem in developing countries, whose demographic growth rate is much higher than in industrialized countries, for socio-cultural and economic reasons, although fertility rates vary dramatically depending on the country. Successful economic development is all the more difficult when the population is growing rapidly.

3.3.5 Human Development Index

In addition to the accounting limits that have been indicated above, GDP/capita and GNI/capita provide only very partial information on what has been called human development since 1990. In an attempt to better understand this difficult to quantify concept, the United Nations Development Program (UNDP) introduced a new indicator called the Human Development Index, or HDI.

It is a composite index, dimensionless, between 0 and 1, determined from three indices quantifying respectively:

- life expectancy at birth, representative of health;
- the rate of adult literacy, representing the ability of an individual to participate in decisions concerning them;
- per capita income (log (GDP/capita) in PPP), representative of the standard of living.

It is generally considered that developed countries have an HDI greater than 0.8.

The graph in Figure 3.2 shows the global distribution of HDI values in 2007. HDI ranges from nearly 1 for most developed countries to about 0.35 for the poorest ones.

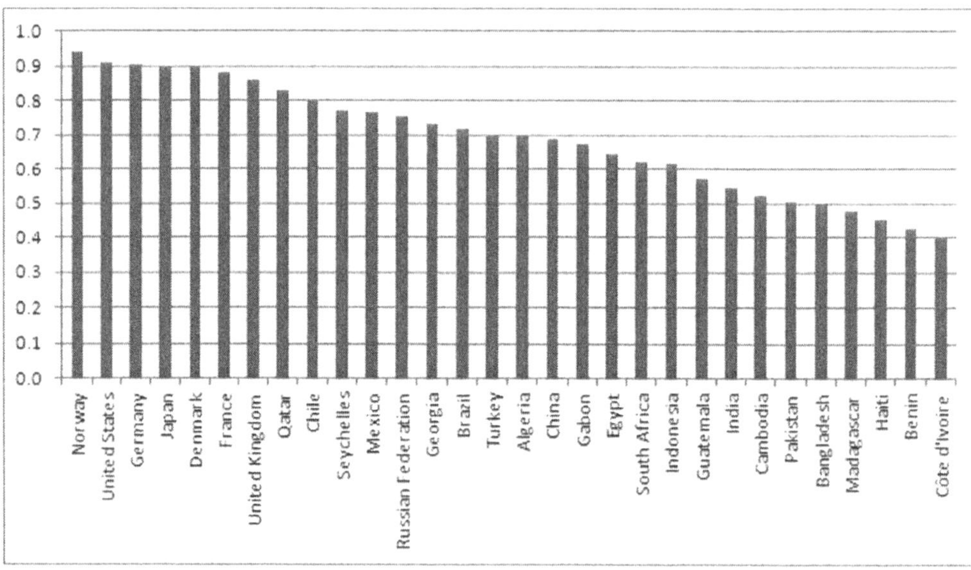

Figure 3.2 Human Development Index in 2007.

Table 3.2 HDI world distribution.

HDI		*Population (millions)*	*Population (%)*
Very high	HDI > 0.9	986.5	14.8
High	0.9 > HDI > 0.8	918.4	13.8
Medium	0.8 > HDI > 0.5	4 380.5	65.7
Low	0.5 > HDI > 0.34	385.1	5.8

Table 3.2 shows the distribution of the human development index in the world.

Although this new index is of great interest, we will not use it in our analyses, both because the available data sets are much shorter than those for GDP or GNI and because its calculation is not fully stabilized yet.

3.4 STRUCTURE OF GDP

GDP being an aggregate indicator, it does not specify the type of economic activity nor its distribution by branch. To access this information, it is necessary to disaggregate the value, which can be done in several ways, by considering areas that can be defined differently depending on the country, the period and the concerns of the analyst.

Historically, the three following traditional sectors have been considered:

- primary sector: all economic activities producing commodities, especially agriculture and mining;
- secondary sector: all economic activities corresponding to the transformation of raw materials into productive assets or consumer goods;
- tertiary sector: all economic activities producing services, such as commerce, banking, insurance, hotels, healthcare, etc.

Of these three sectors, which are hardly still employed as such, only the tertiary is still considered today.

Nowadays, it is often preferred, rather than resorting to the separation between the primary and secondary sectors, to group in the industry all manufacturing and mineral extraction, public works, construction, energy (gas and electricity) and water,

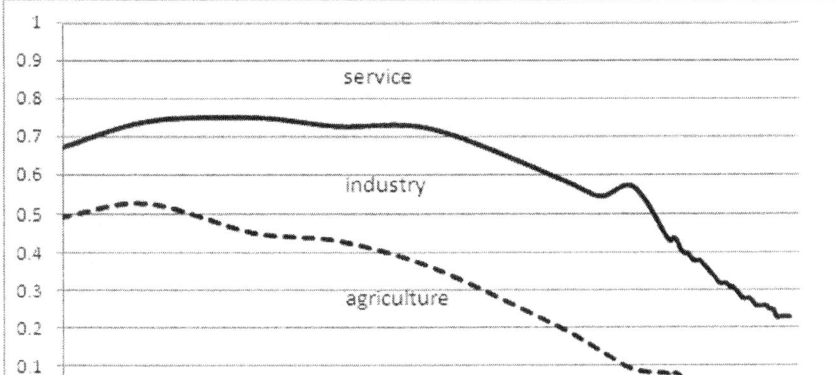

Figure 3.3 Structure of GDP in France.

and to separately identify the agricultural sub-sector. This approach leads to a breakdown of GDP into 3 sections:

- agriculture;
- industry;
- services.

The graph in Figure 3.3 shows the historical evolution of the French GDP structure.

Since 1950, this structure has undergone significant changes: the agricultural sector declined by half while the service sector increased by more than 20 points. In each sector, the respective weights of the different branches (e.g. trade or market services) have also significantly changed.

3.5 ENERGY ACCOUNTING, ENERGY CHAIN

While the accounting of economic activity can only be made, as we have seen, in monetary units, for lack of a standard reference common to all branches, the First Law of Thermodynamics states that all forms of energy are equivalent, which means that it is possible to make an energy accounting in physical terms. In practice, however, some problems arise, as we will see.

In order to be used and meet the needs of consumers, energy needs to be transformed, which is done through the various sectors discussed in Chapter 2: crude oil is refined, yielding among other things gasoline that, burnt in a car engine, is converted into mechanical power, which can satisfy a need for mobility.

This creates an energy chain (see Table 3.3), which connects primary energy, i.e. energy before processing, to the useful energy needed to meet the social or productive requirements (energy services). Between these two extremes, it is customary to distinguish the secondary energy, i.e. energy after processing by the energy branches, and the final energy, which corresponds – but for equipment loss – to the useful energy.

Table 3.3 Energy chain.

PRIMARY ENERGY raw energy before transformation
|
transformation losses (refining, power generation)
|
SECONDARY ENERGY after processing by the energy branches
|
distribution losses
|
FINAL ENERGY delivered to the user
|
end use losses (equipment)
|
USEFUL ENERGY
|
ENERGY SERVICE

Energy is a physical means to address needs of a very diverse nature: power, thermal comfort, mobility, hygiene, health, economic production, etc. These needs may or may not be satisfied depending on whether energy can be obtained at an acceptable cost. There may therefore be a difference between potential demand and actual consumption.

Besides, lifestyle can greatly influence the desired energy service (light, heat, coolness, cold food, travel, information, communication, etc.). The concept of energy sobriety thus complements that of energy efficiency, which is often brought forward.

In those cases where the needs can be met, the amount of spent energy depends on the efficiency of the equipment used: one consumes more energy with an old car than with a new energy-efficient model. Varying amounts of final energy can be needed to meet the same useful energy demand.

Final energy, to be made available to the user, has to be distributed. For example, power is distributed to individuals in low voltage, while high voltage is produced in the plant. Distribution causes losses, which explains the difference between final energy and secondary energy. This is not due to the consumer equipment effectiveness, but rather to the distribution network.

To produce secondary energy, primary energy has to be transformed. For a given form of secondary energy, it is sometimes possible to transform (by various processes) different primary energies. This is particularly the case of electricity, which can be generated in many ways, as we saw in Chapter 2:

- chemical energy (batteries);
- potential energy (hydroelectricity);
- coal or uranium (steam plant);
- gas or oil (steam power plants, gas turbines, generators);
- renewable energy (wind, solar).

The amount of primary energy required to produce a given amount of secondary energy depends on the type of primary energy used and on the efficiency of the process.

Figure 3.4 provides the final energy conversion efficiency of primary energy, which therefore also includes distribution losses, for the 35 OECD countries in 2007. Values range from 51% in Iceland to 82% in Ireland, if we exclude the case of Luxemburg. France's value is 63%, much lower than the European average (70%), which is mainly due to its low electricity conversion. PWR nuclear reactors indeed have an efficiency of about 35%, against 55% for combined cycles, whereas cogeneration is less developed than in other countries.

3.5.1 Accounting problems

Accounting difficulties arise for several reasons:

- firstly, it is often impossible to measure the useful energy. At most, one can measure final energies, and often without knowing the exact breakdown of uses. For example, it is possible to record the electricity consumption of households, but without being able to characterize the share of each item in the consumption (lighting, TV, hot water, etc.).

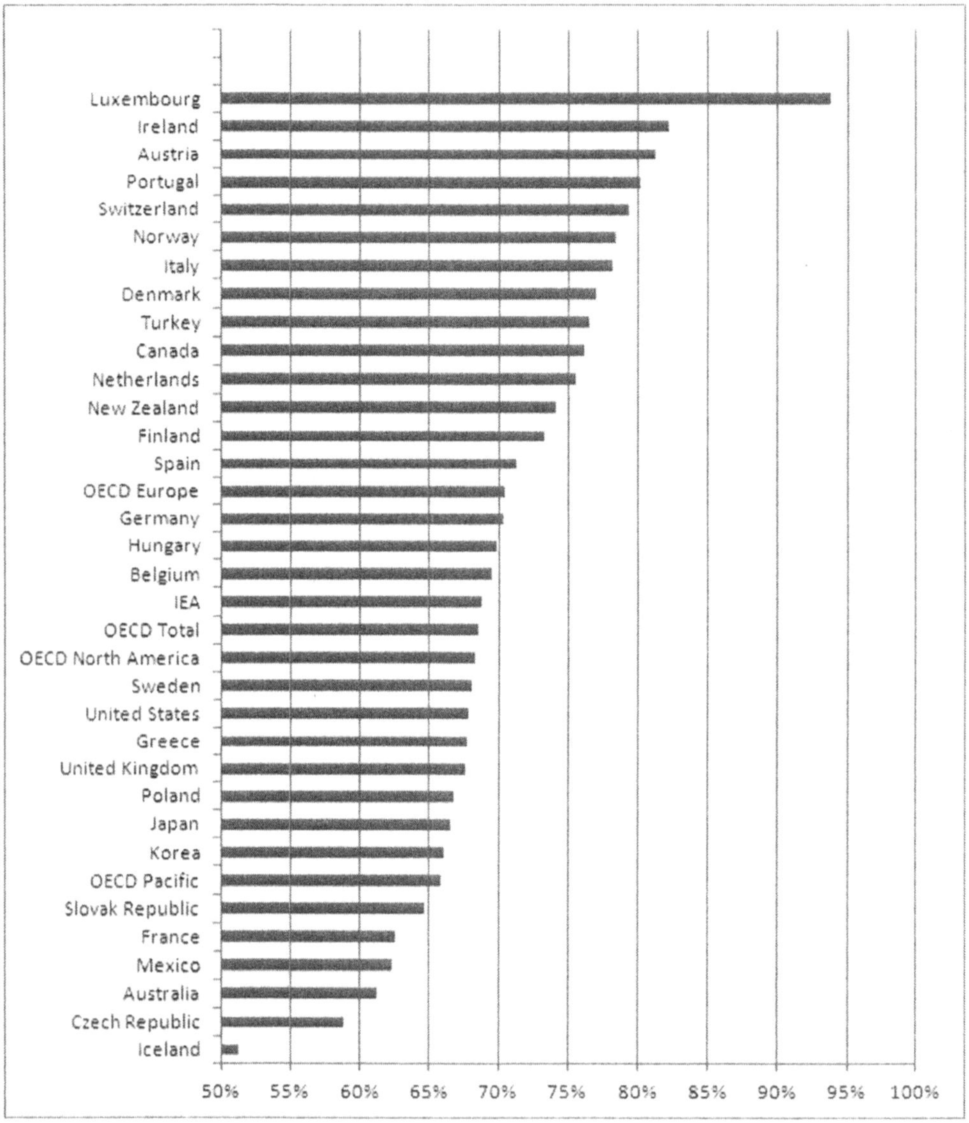

Figure 3.4 Ratio of final energy/primary energy (OECD, 2007).

- only commercial energy can be taken into account precisely, because it is subject to a financial transaction, and must be measured for it. When the share of non-commercial energy is important, energy balances are often very poorly known;
- another problem is that energy is transformed along the energy chain with non-constant conversion efficiencies and losses. Depending on the values (sometimes

arbitrary) one chooses, a particular need for final energy will not correspond to the same primary energy demand.

Determining the primary energy required to produce electricity poses particular problems because of the low conversion efficiency of thermal power plants. There are thus many statistical systems that use different energy equivalences.

Since 2002, France has harmonized its accounting method by adopting the approach common to the IEA, Eurostat and the WEC:

- electricity produced by a nuclear power plant is accounted for using the "primary production equivalent", with a substitution coefficient of 0.2606 toe/MWh;
- all other forms of electricity are accounted for using the "energy content", with a coefficient of 0.086 toe/MWh.

Another example is the combustion efficiency used for firewood: depending on the type of furnace used, it can vary widely, so it is very difficult to match useful energy needs to consumption in terms of quantity of wood.

The result of these difficulties is that the figures may be quite different depending on the estimation methods used. Under these conditions, these figures should be treated with caution. They represent an estimate of energy consumption, but still retain some bias.

One rule in particular must be observed: one should stick to the same accounting system when considering time series, and never mix data from different sources, in order to avoid obtaining purely artificial differences.

Annex 2 provides explanations on the different units which are commonly used to account for energy and their conversion factors.

Primary or final energy are often reported in terms of millions or billions of tonnes of oil equivalent, or, in countries using the British system, in quadrillion Btu (quad):

$$1 \text{ quad} = 10^{15} \text{ Btu} = 25.2 \text{ Mtoe}$$

Barrels of oil equivalent (boe) are also commonly used:

$$1 \text{ toe} = 6.84 \text{ boe.}$$

3.5.2 Flow graphs

To visualize all energy flows entering and leaving the energy system at the regional or the national level, it is customary to use diagrams in which each form of energy is represented at every stage of the energy chain, by an arrow with a thickness proportional to its weight in the total.

In doing so, easy-to-understand flow graphs are obtained, which can be presented at different levels of detail.

The graph in Figure 3.5 shows the flow of energy in the UK in 2011. From left to right there are mainly:

- natural gas;
- coal;

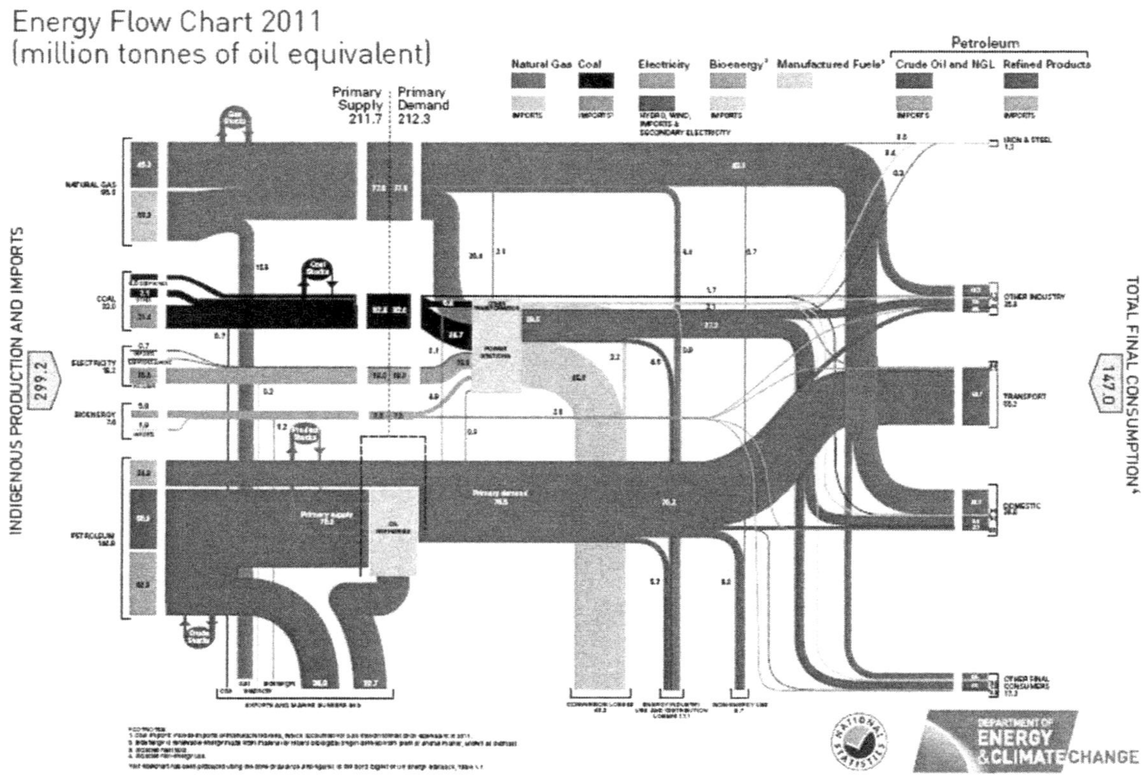

Figure 3.5 Graph of the energy flows in the UK in 2011 (A color and updated version of this figure is available at: https://www.gov.uk/government/organisations/department-of-energy-climate-change/series/energy-flow-charts). [*Source*: UK Department of Energy and Climate Change].

- electricity;
- bio-energy;
- oil.

3.6 ENERGY INTENSITY

The above sections have allowed us to define global indicators representing on the one hand the volume of economic activity, and on the other hand the energy consumption of a country.

Although slightly biased, these indicators are still very useful in practice, in particular to study how economic and energy policies change.

3.6.1 Definition of the energy intensity

We will now try and connect these global indicators, by introducing two new concepts: energy intensity and energy-income elasticity.

Between GNI and GDP, the latter will be preferentially used, as the difference between the two is the international transfer of income, which has a negligible impact on energy consumption.

A particularly interesting indicator is the Energy Intensity (EI) ratio of primary energy consumption to GDP, which measures the energy content of a unit of production.

$$\mathrm{EI} = C_{ENER}/\mathrm{GDP}$$

In general, the energy intensity is expressed relative to primary energy, but it can also be compared to the final energy consumption or to a given source of energy, such as oil, gas or coal. This is known as GDP content in oil, gas or coal.

Energy intensity depends on two main factors:

- GDP structure. We have seen that the GDP can be decomposed into different branches, corresponding to various activities (agriculture, industry, services). The secondary sector consists of industries with high energy content, such as steel, aluminum production, cement, etc. This is not the case for the service sector, with most of the services using only little energy (e.g. electricity for powering office equipment, etc.). Energy intensities of each branch are very different, and the energy intensity of the whole economy depends heavily on the respective share of each one, i.e. the structure of GDP;
- The second factor is the efficiency of the existing energy system. With a given structure of GDP, the energy intensity will be lower if the efficiency of the energy system is higher. Generally speaking, for a given level of development, energy intensities are lower when the country is not endowed with energy resources, and thus encouraged to optimize the operation of the energy system considered as a whole (supply system as well as end-use equipment).

The evolution of energy intensity reflects the structural changes undergone by the national productive system and technological progress in all sectors, especially in terms of energy conservation. This is thus one of the few instruments used to measure the effectiveness of energy policies implemented by countries.

Note however that the energy intensity of two countries can only be compared in relation to a common reference standard value of GDP in national currency. As long as the exchange rate fluctuates, energy intensities vary without this evolution having any actual meaning. Here we find one of the difficulties regarding international comparisons.

Energy intensities are generally expressed in kilograms of oil equivalent per dollar, or, in the US, in kBtu per dollar.

3.6.2 World energy intensities

The graph in Figure 3.6 shows the values of energy intensity in 2010 in a large number of countries (the left y-axis scale is in kilograms of oil equivalent per PPP dollar [in 2005], and the right y-axis scale in kBtu per PPP dollar [in 2005], with the x-axis scale in kilograms of oil equivalent per capita), including those that are the subject of case studies presented in Chapter 7.

We can see that in most countries, energy intensity is less than 0.3 koe/dollar. In some cases, much higher values can however be identified.

3.6.3 Income elasticity

Economists and governments are often more interested in flows than stocks, and even more in flow trends than in flows. Indeed, their main concern is often economic

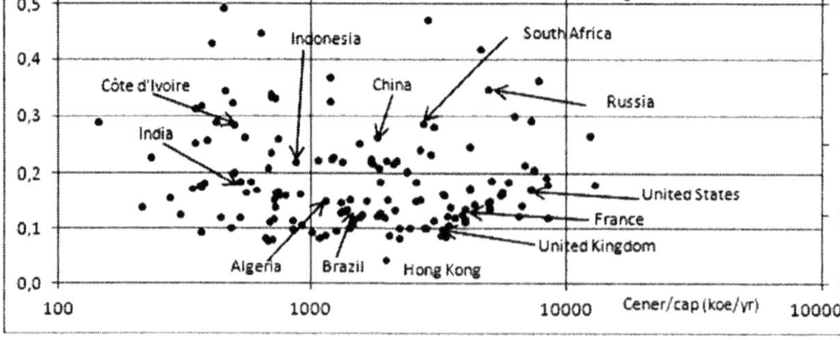

Figure 3.6 World energy intensities.

growth (or its opposite, the recession). The absolute value of GDP is considerably less cited than its growth rate.

In many countries, the rate of economic growth has become the yardstick by which governments are judged. Given the fundamental role of energy as a factor of growth, it is important to be able to predict the amount of energy that will be needed to achieve the economic goals that are set. This is all the more necessary considering that implementation delays are very long in the energy sector so that forecast errors cannot be quickly corrected.

In general, elasticity is a classic economic concept to connect the variation of a quantity that one seeks to understand to the evolution of an indicator that is assumed to be known.

For this reason, in addition to energy intensity, it is customary to examine another indicator: the energy demand–income elasticity, defined by the relation:

$$e = \frac{\text{rate of } C_{ENER} \text{ growth}}{\text{rate of GDP growth}}$$

Since the growth rate can be expressed as logarithmic derivatives, this represents, between energy consumption and GDP, a relation of the type:

$C_{ENER} = k\ \mathbf{GDP}^{e}$, where k is a constant.

The concept of energy demand-income elasticity was especially popular due to the fact that it remained relatively stable for many years before the first oil shock, its value being close to 1 or slightly lower in the case of developed countries, and ranging between 1.5 and 2 in the case of developing countries in their industrial takeoff.

Forecasting energy demand–income elasticity then allowed for easily estimating the energy needs over the long term, as soon as a scenario of economic growth was chosen by decision-makers.

As with energy intensity, the higher energy demand-income elasticity in developing countries is mainly due to structural factors (infrastructure development, energy intensive industrial development).

Using the concept of energy demand-income elasticity in predicting long-term needs implicitly reflected the certainty of a deterministic and strong link between GDP growth and energy consumption growth.

We will see later that this concept has since been challenged.

3.6.3.1 Energy hill

Historically, depending on a country's stage of development, its energy intensity and income elasticity increase during the industrialization phase and peak, and then decrease as it enters the post-industrial phase. This change has given rise to the concept of "energy hill" represented by the diagram in Figure 3.7.

A number of forecasters have speculated that sound energy policies could allow countries to dig a "tunnel" in the energy hill in order to avoid the phase of increasing energy intensities corresponding to their industrialization. We will see later that if the maxima certainly tend to decrease due to technological advances among other factors, they always exist.

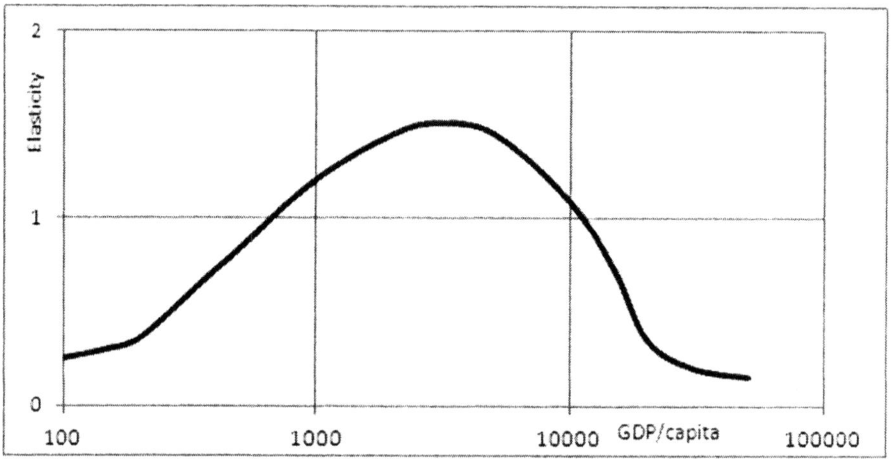

Figure 3.7 Sketch of the energy hill.

3.6.4 Relationship between growth rates

We have seen that the evolution of the energy intensity of a country depends on the structure of its GDP and the effectiveness of the national energy system and makes it possible to assess the effect of energy policies implemented by countries. We will now study the relations linking the rates of change of GDP, energy consumption and energy intensity.

By definition, the energy intensity is equal to $EI = \frac{C_{ENER}}{GDP}$

Let us call:

- p the growth rate of GDP;
- c the growth rate of energy consumption;
- i the growth rate of energy intensity;
- e the energy-income elasticity.

We have by definition:

$$p = \frac{d\,GDP}{GDP}$$

$$c = \frac{d\,C_{ENER}}{C_{ENER}}$$

$$e = c/p$$

$$i = \frac{d\,(C_{ENER}/GDP)}{C_{ENER}/GDP} = c - p$$

Thus: $$\boxed{i = p(e-1)} \quad (3.1)$$

or: $$\boxed{e = 1 + \frac{i}{p}} \quad (3.2)$$

$$\boxed{c = p + i} \quad (3.3)$$

The arithmetic expressions written above require a number of comments:

- i can be considered as a relatively stable technical coefficient in the short term, and varying slightly in the medium term due to the effects of structure and technological change;
- equation (3.2) explains the wide variation in elasticity in the period after the oil shocks: even if i is assumed to be relatively stable when p is close to 0, which was the case between 1979 and 1982, e is able to take almost all possible values, without this having any significance regarding the evolution of the energy content of the GDP in the short term;
- one can define e and i in reference to the total energy consumption, primary or final, which corresponds to the classical definition, or use a particular source of energy such as oil (known as the oil content of GDP: in this case, one uses the notations EI_p, i_p, e_p and c_p);
- if one seeks to determine the evolution of the energy or oil growth rates, values of concern are c or c_p, and equation (3) is the most reliable and easiest to use. It does not involve the energy income elasticity, which, as we have seen, is difficult to use and can be a source of error;
- if one refers the energy intensity to final and not primary consumption, it does not take into account the efficiency of the supply system, but only that of energy use equipment.

In international comparisons, one has to take into account that i can incorporate assumptions about exchange rates that could bias the results.

In conclusion, during periods of high variation of the GDP growth rate, the use of energy-income elasticity should be avoided since it loses its meaning, and the evolution of energy consumption should be calculated from the very simple relationship: $c = p + i$.

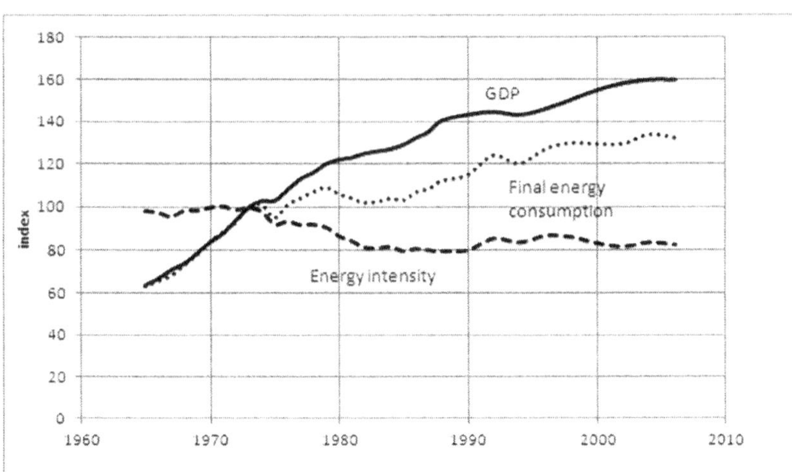

Figure 3.8 Energy intensity in France.

3.6.5 GDP and energy in France

The graph in Figure 3.8 shows the changes between 1965 and 2007, of GDP, energy consumption and final energy intensity in France, offset against an index of 100 in 1973.

It reflects structural changes and energy efficiency policies that have been adopted during this period: between 1973 and 2007, the GDP grew in value by 60%, while the energy consumption increased by 32%. During this period the energy intensity decreased by almost 20%.

Whereas energy consumption increased faster than the GDP before the first oil shock, the opposite happened afterwards.

3.6.6 Energy intensity changes of OECD countries

An examination of the changes in energy intensity of OECD countries shown in the graph in Figure 3.9 is also very telling, although it reveals disparities between countries (the left y-axis scale is in kilograms of oil equivalent per dollar [in 2000], and the right y-axis scale in kBtu per dollar [in 2000]).

Note that the only country whose energy intensity has increased significantly over the last decades is Iceland [Solutions, 2012]. This phenomenon can be explained by the specificity of Iceland's energy balance. Let us recall that in section 3.5 we saw that Iceland is the country where the final energy/primary energy conversion efficiency is the lowest (51%) in the OECD. This is partly explained by the existence of a heavy industrial aluminum production that consumes a lot of electricity (two-thirds of domestic consumption). Iceland has also undertaken a major energy transition since the 1960s from fossil fuels (coal, oil) to renewable energy (geothermal, hydro), which represent more than 80% of the national mix, and aims at reducing its CO_2 emissions to zero by 2050.

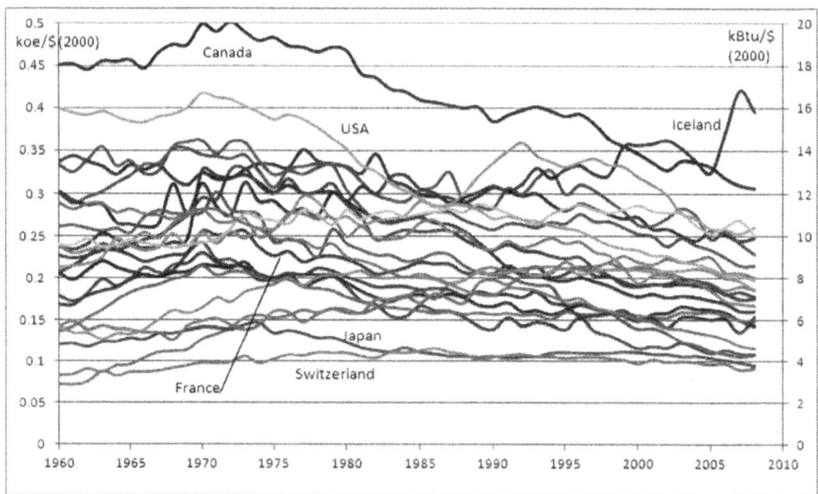

Figure 3.9 Energy intensity of OECD countries.

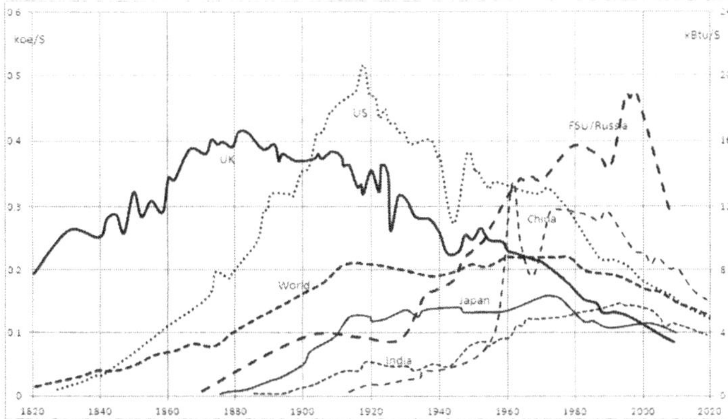

Figure 3.10 Long-term changes in energy intensities.

3.6.7 Convergence of world energy intensity

In the previous sections we mentioned the many challenges that arise when trying to obtain consistent energy and economic indicators to compare the situations in countries of different economic levels and during different time periods.

Based on rigorous methodological work carried out by specialists, BP economists studied long-term historical trends of energy intensity of a number of countries (United Kingdom, United States, Japan, Russia (including former Soviet Union), China and India [Rühl et al., 2012], Figure 3.10)

Except for a few disturbances linked to specific historical events, all display a similar bell curve. The slope on the left, which corresponds to the phase of industrialization, is steeper than that on the right-hand side, which reflects the development of the service sector and the post-industrial phase, taking place when the infrastructure has been built.

Note that the centrally planned economies have generally had higher energy intensities than the others, because of the distortions usually induced by planning compared to a purely economic allocation of resources.

Finally, one of the most remarkable points highlighted by this work is the global convergence of the evolution of energy intensities, a recent phenomenon, observed since the beginning of the century: energy intensities in most countries are now between 0.1 and 0.2 koe/dollar in 2010, and are all heading downwards (about 2% per year).

3.6.8 Changes in the electricity content of GDP

The graph in Figure 3.11 shows the evolution of OECD GDP electricity content, but with a much more expanded scale than the previous one. Although quite different from that of energy intensities, it also looks like a bell curve, with a sharp increase from 1970 to 1980, followed by a stabilization for fifteen years and then a steady decrease.

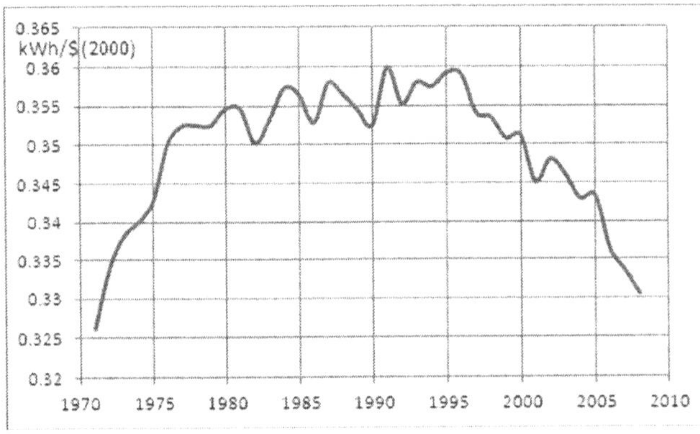

Figure 3.11 OECD GDP electricity content.

Globally, electricity consumption is growing much more strongly than energy, many countries still having only very limited infrastructure for electricity, especially in rural areas. This form of energy being immediately correlated to a rising standard of living, major development programs are devoted to it, which explains this strong growth. As a result, the growth of electricity consumption is roughly equal to that of GDP, and thus GDP electricity content remains almost constant.

3.7 INDICATORS FOR CO_2 EMISSIONS

In section 8.2 we will see that carbon dioxide is the gas that is by far the most responsible for the greenhouse effect increase and that 74% of CO_2 emissions are today due to burning hydrocarbons and coal. As a result, more and more people want to have indicators related to those emissions.

While energy indicators are, as we have seen, based on measured physical quantities that are subject to market transactions, the amount of CO_2 emitted is seldom measured. It is estimated on the basis of conventional emissions values. In France, for instance, these values are defined in a government circular on greenhouse gas emissions accounting [Circulaire, 2002].

When estimating CO_2 emissions, at a national or global level, from the combustion of fossil fuels or biomass, one encounters a number of methodological problems:

- Firstly, the amounts of recorded primary fuels include non-energy uses that do not result in CO_2 emissions (such as the use of oil as feedstock for petrochemicals, and bitumen). Thus, strictly speaking, one must deduct non-energy uses from the primary balance;
- Secondly, the amount of emitted CO_2 depends on the burnt fuels, whose diversity disappears from primary balances, where one finds only three categories: coal, oil and natural gas. Equivalence ratios, often varying from one country to another, are used for conversions, and as one only gets aggregated values, one has to adopt an average estimation method. This can be even more delicate as

primary energy is usually expressed in a common unit, usually in tonne of oil equivalent;

- Finally, in practice, many chemical reactions are incomplete, due to the onset of dissociation at high temperature. In this case, the involved energies may differ significantly from theoretical values corresponding to complete reactions.

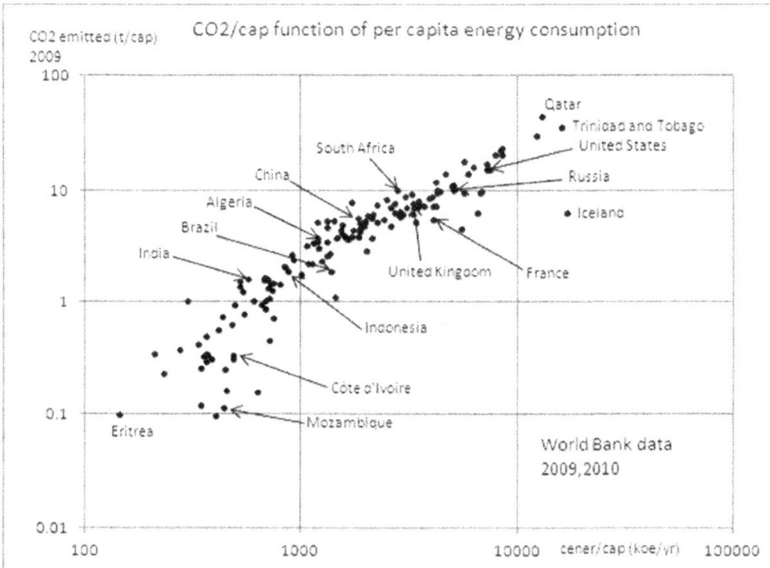

Figure 3.12 CO_2 emissions per capita in terms of per capita energy consumption.

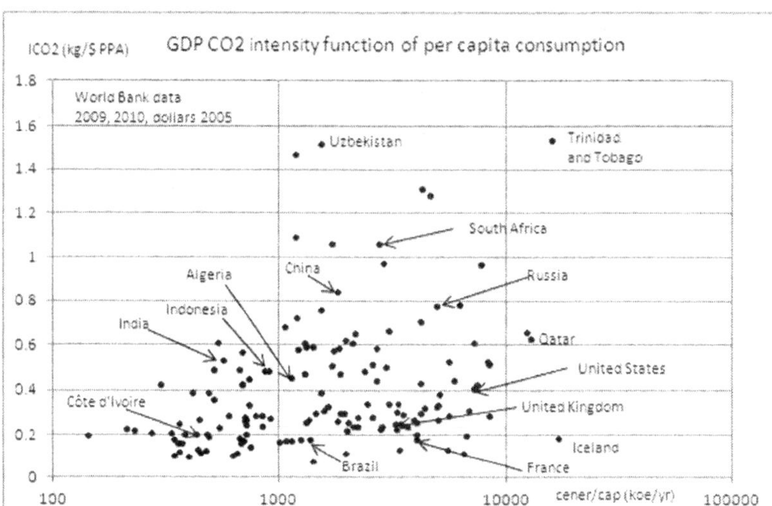

Figure 3.13 CO_2 intensity of GDP depending on per capita energy consumption.

National and international statistical offices therefore take appropriate precautions when estimating emissions, expressed in metric tons, which provides a first indicator. As for GDP or energy consumption, it is often useful to relate these quantities to the population, which provides a second indicator, per capita emissions. Finally, it is common to link emissions to GDP, by defining the CO_2 intensity of GDP, or the income elasticity of GDP in CO_2 emissions.

Figure 3.12 gives the values of per capita CO_2 emissions depending on the per capita energy consumption for 150 countries. It clearly shows that these two indicators are highly correlated for most countries, with a few exceptions such as Iceland, which we have already pointed out in section 3.6.6. This country, whose per capita energy consumption is very high, emits far less CO_2 than other countries of similar levels of economic development because more than 80% of its energy is provided by renewable energy.

Given the purpose of this book, more dedicated to energy issues than environmental ones, it seems unnecessary to transpose all its analyses on per capita energy consumption or energy intensity to express them related to CO_2 emissions. The conclusions would be the same except for a very small number of countries.

We will therefore only provide Figure 3.13, which gives the intensity of GDP in CO_2 emissions depending on the per capita energy consumption for about 150 countries, those examined in Chapter 7 being explicitly positioned.

3.8 ENERGY MARKETS

This section focuses on the international exchange of energy, distinguishing oil, natural gas and coal, which are governed by different logics.

3.8.1 International trade

International trade in energy was born and has developed because of the existence of geographical regions where the (non-renewable) energy deposits are distinct and remote from centers of consumption.

After increasing from 1960 to 1973 by 10% per year on average, in 1973 the international energy trade accounted for a volume of 2.2 billion toe (Gtoe), representing 44% of the global consumption of primary energy. Note that these figures include trade in refined products. Without them, the total trade was 1.75 Gtoe, representing 30% of global primary energy consumption.

At that time, international trade therefore accounted for an essential component of the global energy system, its macroeconomic impact starting to become significant. It accounted for approximately 10% of the total world trade in goods, a share that would more than double to reach nearly 25% in 1981, before dropping to about 18% in 1985 and 15% in 2006.

In 1973, the share of oil in international energy exchanges was by far the most significant (70% for crude oil and 20.3% for refined products), with coal and natural gas playing a marginal role (6.2% and 3.4% respectively).

These figures may be compared to the global commercial energy balance, where the respective shares of these three sources of energy were 55.6%, 31.1% and 20.8%.

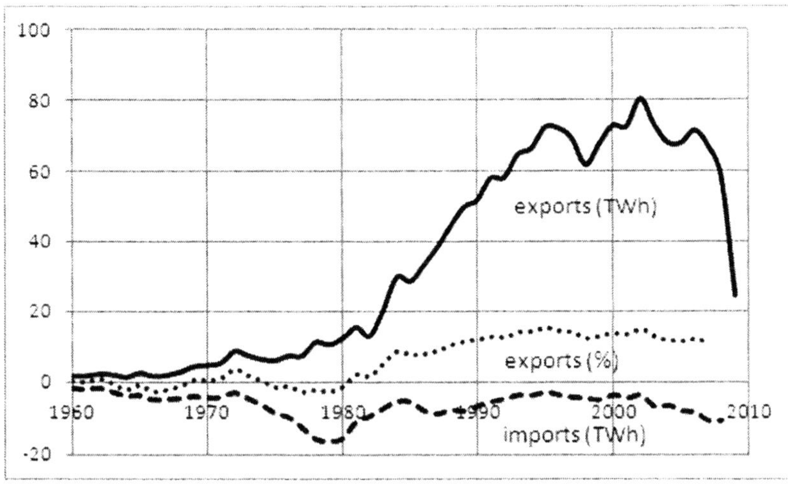

Figure 3.14 International trade of electricity in France.

In other words, international trade in energy represented highly variable shares from one energy source to another: 54.7% for crude oil (70.6% taking into account re-exports of refined products), against only 8.8% for coal and 7% for gas.

This very striking contrast must be kept in mind, as it explains in large part why energy prices have been and remain dominated by the price of oil.

In 1984, the global energy trade had decreased by only 10% in volume compared to 1973 (2 billion toe), but its share in primary energy consumption fell from 44% to 32%. In 2011, it accounted for about 30%.

In 1960, international trade in electricity was relatively low, with an almost negligible net worth. Since then, its growth has been steady (10% per year until 1973 and 5% per year during the 20 years that followed, reaching 250 TWh in 1996).

The graph in Figure 3.14 shows the international trade in electricity in France, in TWh and as a percentage of production. The simultaneous existence of imports and exports is explained by the ongoing exchange taking place at the borders to stabilize the European network.

The middle curve represents the share of the exported net production, that is to say the balance of exports and imports related to total production. Note the sharp drop in exports that took place in 2009, which is linked to:

- reduced availability of nuclear power plants;
- growth of the national electricity consumption;
- external demand decrease due to the economic recession.

3.8.2 Oil market

Oil can be considered as a rather standardized raw material, regardless of its origin, which could be exchanged at a single world price. This type of good is called a commodity. Strictly speaking, this is not entirely true, but refining makes it possible to distinguish the properties of various crudes from those of refined products.

Crudes being largely interchangeable, the importer is not necessarily directly related to the exporter. An importer can choose his/her supplier according to the law of the market at the best price at a given moment. The flexibility of the distribution and the low share of the cost of transport in the price allow for the oil to be delivered to wherever the importer might wish.

This fundamental characteristic greatly accelerated the development of the competitive market of crude oil over the past forty years, replacing the long-term contracts that existed prior to the first oil shock. Indeed, at that time, the oil market consisted of two unequal components:

- long-term contracts established between large multinational companies and producing states which accounted for the bulk of transactions;
- a competitive spot market, which had been established by the majors as a marginal adjustment mechanism, particularly to ensure the supply/demand adequacy in the market for refined products in Northern Europe. In 1973/74, only 5% of crude oil was traded through this market.

Since the mid-1990s, oil pricing has been based on a mechanism that takes into account technical cost structures and trading on the spot market, with several drawbacks, including the risk of speculative attacks.

3.8.2.1 *Evolution of prices*

The spot market has played an increasing role in the transactions of crude oil, its share in this trade occasionally reaching 50%, or even 70%. The end of trust between producers and consumers, necessary in long-term contracts, has led both buyers and sellers to prefer to go to a market reflecting the laws of supply and demand. The emergence of a crude spot market led OPEC to lose a good part of its initiative in setting oil prices.

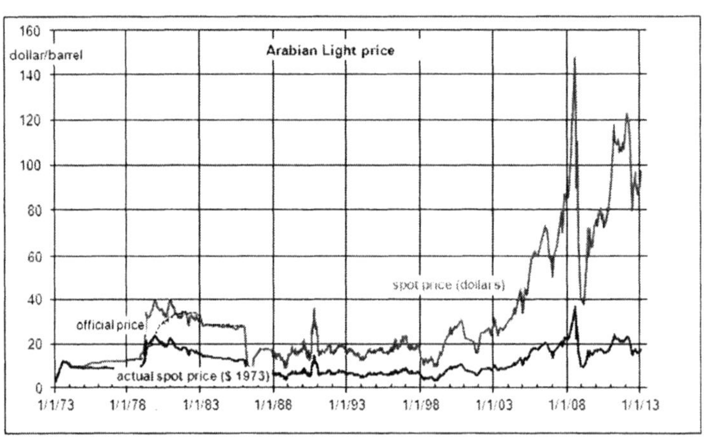

Figure 3.15 Price of Arabian Light.

One of the consequences of the loss of control of the pricing mechanism by large companies and states was the appearance of speculative "fevers", reflected by sudden price spikes without any physical shortage (see Figure 3.15). In particular, the second oil crisis was initially triggered by a simple over-pricing in the spot market compared to official (OPEC) prices, following fears of a resurgence of consumer shortages due to the events in Iran in 1979 and the outbreak of the Iran-Iraq war. Similarly, the events of the first Gulf War confirmed this trend: in 1990 there was a new surge in prices for purely speculative reasons, while there was no shortage, and even as OPEC had agreed to produce additional amounts necessary to compensate the embargo on Iraqi and Kuwaiti crude.

The same phenomenon was demonstrated in 2011 when Libyan oil was not exported anymore due to the civil war, with domestic production falling by 71% compared to 2010.

The soaring oil prices in 2008 were due to many factors, firstly the strong demand from Asia, including China, and also a decrease in production in Venezuela, following the oil sector strikes in this country, and in Iraq because of the war.

3.8.2.2 *Formula pricing contract*

The principle of formula pricing contract is as follows: the price of oil exported is fixed relatively to the import spot price, taking into account the cost of refining, reflecting differences in quality of both crudes and incurred freight costs [CNUCED, 2012].

For the buyer, this guarantees that the ex-refinery price will be comparable even for different crude qualities. For the seller, this guarantees that oil revenues will return to him.

This somehow indexes prices to the spot market, which helped to establish new long-term contracts between exporters and importers.

3.8.2.3 *Internationalization of the oil market*

This change had the effect of further internationalizing oil markets, distance no longer being a barrier to trade. In addition, the global crude market has become more differentiated, thereby reflecting the variations in refined products markets.

Indeed, the trend in industrialized countries to shift the demand towards light cuts means that refining costs (for cracking, reforming, deep conversion) are higher than elsewhere. With refiners impacting costs on products to maintain their profit margin ("netback"), there is a strong correlation, at least in the medium term, between price variations of crude oil and of refined products.

3.8.3 Natural gas market

Natural gas can also be seen as a commodity, but it is nowhere near as versatile as oil because of the weight of transport, storage and distribution infrastructures.

As we saw in Chapter 2, the Liquefied Natural Gas (LNG) chain presupposes the set-up of liquefaction terminals in the exporting country and regasification terminals in the importing country, while pipeline construction requires the installation of a network of pipes, very expensive especially if a sea must be crossed.

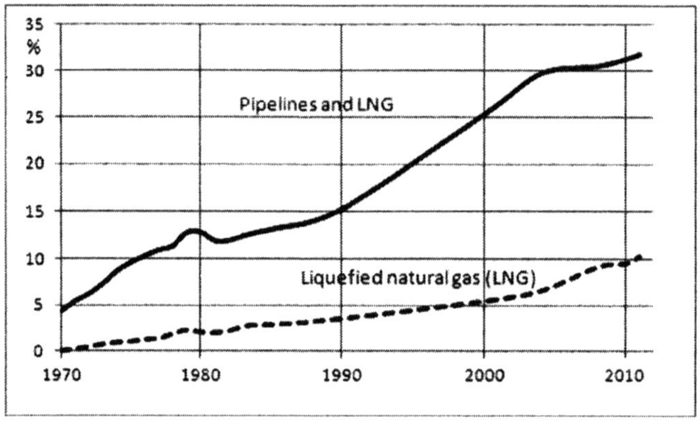

Figure 3.16 Share of international trade in the total production of natural gas.

Unlike oil, the development of such infrastructure was mainly carried out within the framework of long-term contracts linking producers and consumers, called "Take or Pay", the buyer committing in advance to acquire given volumes of gas regardless of immediate needs.

Moreover, vulnerability to accidents and attacks of all gas infrastructures is much higher than that of oil installations.

All these reasons explain why the share of gas in the international energy trade has long remained relatively low. It is facing strong growth however, particularly for combined cycle power plants designed to meet the increasing demand for electricity.

The graph in Figure 3.16, compiled from Cedigaz data, shows the evolution of international trade as a percentage of total production, the upper curve representing both the LNG and pipeline systems, and the lower one LNG alone.

3.8.4 Coal market

The term "coal" covers varied mineral properties, both in terms of heating value, as we saw in Chapter 2 (27 MJ/kg for some lignites, 37 MJ/kg and above for coal) and in terms of impurities (sulfur, nitrogen compounds, etc.). In these conditions, different coals are not interchangeable nor it is possible to define a universal price for coal.

Sales contracts are contracts that specify not only the unit costs and quantities, but also the qualities of coals.

In addition, the low heating value of coal and its solid nature mean that the cost of transportation is relatively much higher than that of oil, so that the user delivered price (cost insurance freight or CIF) is much more sensitive to the remoteness of production than for oil.

Coal also requires infrastructure for its delivery and storage, which is a significant expensive prerequisite for the use this energy source. Furthermore, in industrialized countries, the only ones that buy coal on the international market, it is used almost exclusively for electricity generation and does not therefore serve as a competitor to oil, but to nuclear energy.

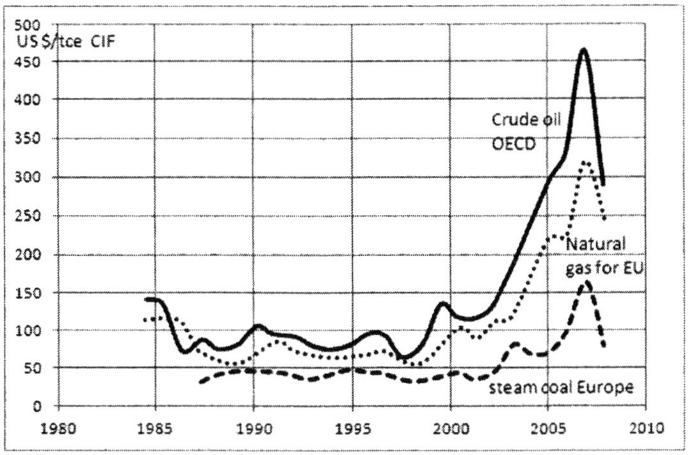

Figure 3.17 Comparison of fuel prices.

3.8.5 Conclusion on the energy markets

As we have seen, the various energy markets are in fact very different from one another: coal and gas exchanges are governed by supply contracts negotiated on a bilateral basis, at least in the context of medium-term policy of both sides, or genuine partnership, while oil, which is now partially released from the grip of major international companies, is traded more on a free and competitive market, and thus follows the changes in supply and demand. Given its relative share in the international energy trade, the price of oil is still the energy reference price (see Figure 3.17).

When the supply is in relative excess to demand, the price of coal can be considered relatively independent from that of oil. It can move between a floor price based on costs of production and transport, and a ceiling price determined by potential offering, risks of cartelization being virtually excluded. The floor price is around $30 per metric ton CIF, equivalent to $3/boe. The historical evolution of prices, however, shows that the price of steam coal follows that of oil (Figure 3.17).

One of the main advantages of gas is its cleanliness, especially in terms of CO_2, sulfur and nitrogen oxides emissions. Other than that, the infrastructure costs encumber its price. The price of natural gas, however, can still be considered indexed to oil or fuel oil price, the underlying reason being that the main exporters (who usually also export oil) avoid putting themselves in direct competition.

That being said, we have seen that these two hydrocarbons have very different commercial characteristics. While one is likely to accommodate a competitive spot market, the other calls for a partnership in the long term: their cost structures are radically different.

Increasingly, some voices in consumer countries are arising to challenge this indexation, on behalf of the tenets of liberalism, considering that users will be the first beneficiaries of a competitive market [OFEN, 2005]. They wish in particular that long-term contracts be renegotiated for shorter periods to allow the entry of new players.

The natural gas market has indeed started to liberalize, first in the United States and the United Kingdom, and the European Union intends to continue in this direction.

REFERENCES

Circulaire du 15 avril 2002 relative aux modalités de contrôle par l'inspection des installations classées des bilans annuels des émissions de gaz à effet de serre, http://aida.ineris.fr/textes/circulaires/text4222.htm

CNUCED, Pétrole, [Online] Available from: http://r0.unctad.org/infocomm/francais/petrole/prix.htm/[Accessed 09/20/2012]

OFEN, Couplage du prix du gaz naturel avec celui du pétrole, Office fédéral de l'énergie, Berne, 2005, [Online] Available from: http://www.bfe.admin.ch/themen/00486/00487/index.html?lang = fr&dossier_id = 00743/[Accessed 01/18/2013]

Rühl Ch., Appleby P., Fennema J., Naumov A., Schaffer M., Economic Development and the Demand for Energy: A Historical Perspective on the Next 20 Years, BP 2012, http://www.bp.com/liveassets/bp_internet/globalbp/STAGING/global_assets/downloads/R/reports_and_publications_economic_development_demand_for_energy.pdf

Solutions, Energy Solutions in Iceland, [Online] Available from: http://www.thesolutionsjournal.com/node/637/[Accessed 12/26/2012]

FURTHER READING

Ayoub A., Percebois J., Ed. Pétrole: Marchés et Stratégies, Economica, Paris, 1987.

Bénassy-Quéré A., Coeuré B., Jacquet P., Pisani-Ferry J., Economic Policy: Theory and Practice, Oxford University Press, New York, 2010.

Boussena S., Prix du pétrole et stratégies de l'OPEP, Revue de l'Énergie, Paris, n° 456, février 1994, CPDP. Bulletins du Comité Professionnel Du Pétrole, Paris.

Chevalier J.-M., Barbet Ph., Benzoni L., Économie de l'Énergie, Presses de la Fondation nationale des Sciences Politiques et Dalloz, Paris, 1986.

Fericelli J., Lesourd J.B., Énergie: Modélisation et Économétrie, Economica, Paris, 1985.

Galbraith J.K., Salinger N., Tout savoir ou presque sur l'économie, Editions du Seuil, 1978.

Giraud A., Boy de la Tour X., Géopolitique du pétrole et du gaz, Technip, Paris, 1987.

Hansen J.P., Percebois J., Énergie, économie et politiques, De Boeck, Bruxelles, 2010, ISBN 978-2-8041-6177-4.

Mabro R., Ed. The 1986 Oil Price Crisis, Economic effects and policy responses, Oxford University Press, 1988.

Martin J.-M., L'Économie Mondiale de l'Énergie, Repères, La Découverte, Paris, 1990.

Percebois J., Économie de l'Énergie, Economica, 1989.

Rad Serecht F., Le marché pétrolier international: ruptures et nouvelles configurations, la Documentation Française, 1985.

Chapter 4

The impact of energy on economic development

4.1 PRESENTATION

This chapter presents the historical impact of energy on economic development in quite general terms.

After showing the strong correlation around the world between energy consumption and economic production, it provides some insights into the historical importance of energy as a factor of development: coal was the "engine" of the Industrial Revolution and numerous work/energy substitutions took place when relevant technologies emerged. The low cost of energy has led to a significant decrease in the cost of transport, allowing for the development of trade of goods and the mobility of people.

This chapter then presents the different energy uses and explains what underlies the formation of demand, before concluding with a section analyzing the macroeconomic impact of energy in modern societies (investment, balance of payments, consumption, revenues of states).

4.2 ENERGY CONSUMPTION AND INCOME LEVEL

The graph in Figure 4.1, expressed in log-log coordinates, shows the distribution of GDP per capita relative to per capita energy consumption for many countries in 2010, GDP being expressed in 2005 dollars and in Purchasing Power Parity (PPP).

As shown in Figure 3.6 on energy intensities, we have indicated in this diagram the position of a number of countries, including those that are the subject of case studies presented in Chapter 7.

Even if the point cloud shows some dispersion, this graph clearly demonstrates that the higher the per capita income, the higher the per capita energy consumption. In other words, economic activity feeds, among other things, on energy.

4.2.1 Factors of development

Energy appears as a factor of production, along with labor, capital and natural resources.

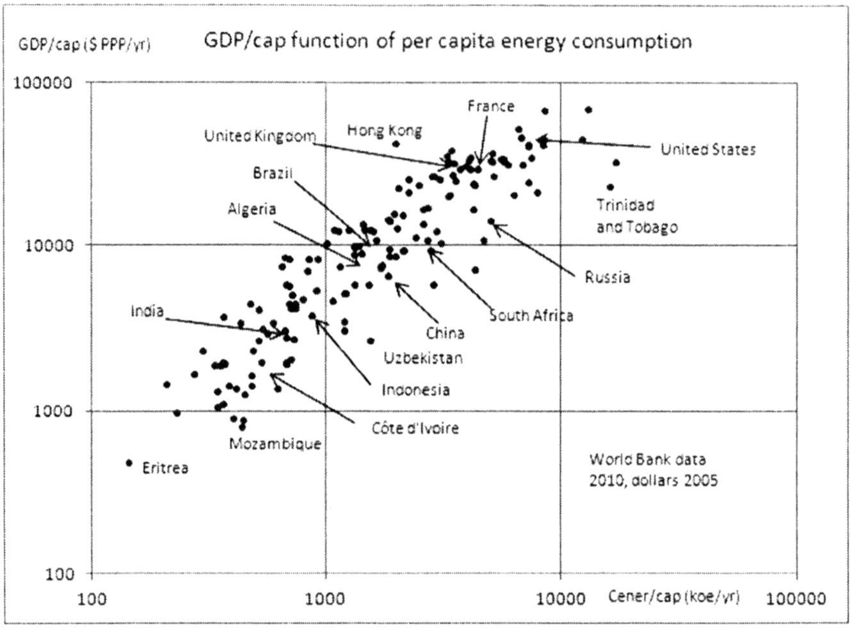

Figure 4.1 GDP per capita function of energy consumption per capita.

How these factors combine in the economy depends greatly on the context (including the type of economic system), but to a varying extent, they are all needed.

- capital is required to purchase production equipment. It also integrates existing assets, such as industrial equipment, agricultural land, etc.;
- labor of a suitable qualification is required to operate the equipment, to manage, to sell, etc.;
- natural resources are generally a structuring feature. They include mineral resources such as water, soil quality, climate, etc.

Substitutions are possible between these different factors, including between energy and labor.

The capital-labor substitution is a phenomenon closely studied by economists, due to its impact on employment and also because it seems that it can induce effects on the evolution of energy intensity.

Energy and capital are essentially complementary, since energy is usable only through equipment. However, substitutions are also possible when a capital injection reduces the need for useful energy. It is one of the goals of energy efficiency and demand management, and the results obtained during the last forty years have proved the reality of this substitution potential.

A brief history of the evolution of energy in our societies will highlight the profound influence of this factor on our daily socio-economic environment.

4.2.2 Evolution of energy consumption

Energy in its current form is a newcomer to the landscape of human civilization. For millennia, man has only exploited renewable energy, including energy from animals. Powers at his disposal remained minimal, limiting his ambition and his ability to transform the world. The great empires of the past all based their power on forced labor or slavery, because they lacked sufficiently concentrated energy, etc. Until the advent of the steam engine, the most powerful monarch in the world could not move faster than a horse.

As shown schematically in the diagram in Figure 4.2, energy consumed by humanity has evolved through the ages, both in volume and composition [Cook, 1971].

It is obvious that when the per capita energy consumption is very low, its growth substantially increases the wellbeing and standard of living, whereas it is not assured that past a certain threshold, this increase is still socially beneficial.

The graph in Figure 4.3 indeed shows that life expectancy increases sharply with consumption when it is less than 1 or 2 toe/capita, but stabilizes beyond this value.

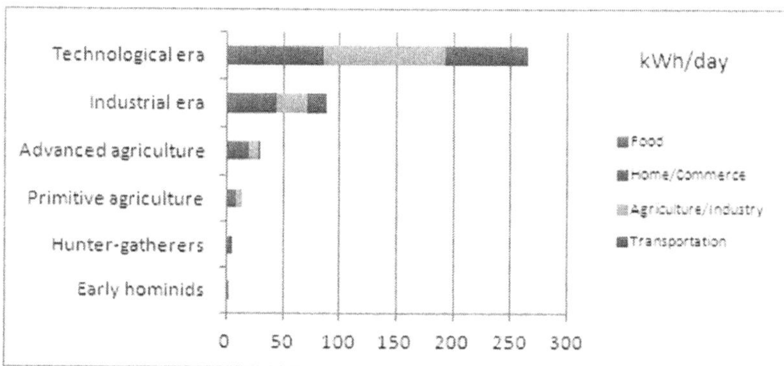

Figure 4.2 Evolution of energy consumption (kWh/day).

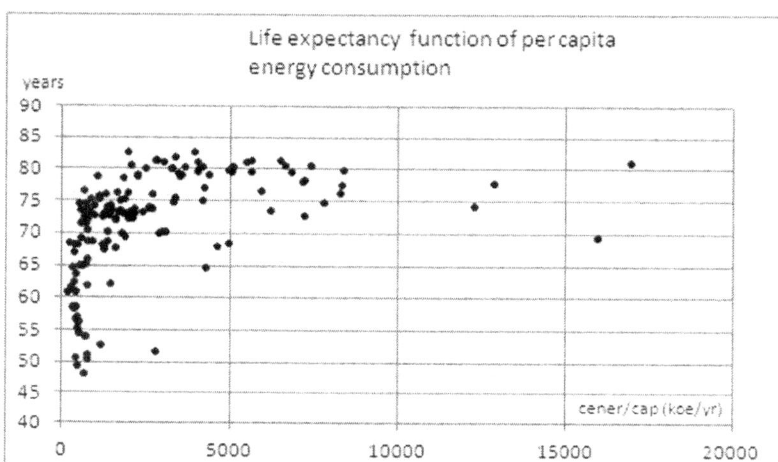

Figure 4.3 Life expectancy and energy consumption in 2010.

Since the industrial revolution, the global energy consumption has continued to grow, as shown by Figure 4.4, whose left axis represents Gtoe and the right quadrillion Btu (quad). It has been multiplied by a factor of 100 over the last 150 years, growing at a faster than exponential rate until the first oil shock.

Meanwhile, the available power unit was multiplied by more than one million between that corresponding to a horse and the thrust created by a modern rocket (Figure 4.5).

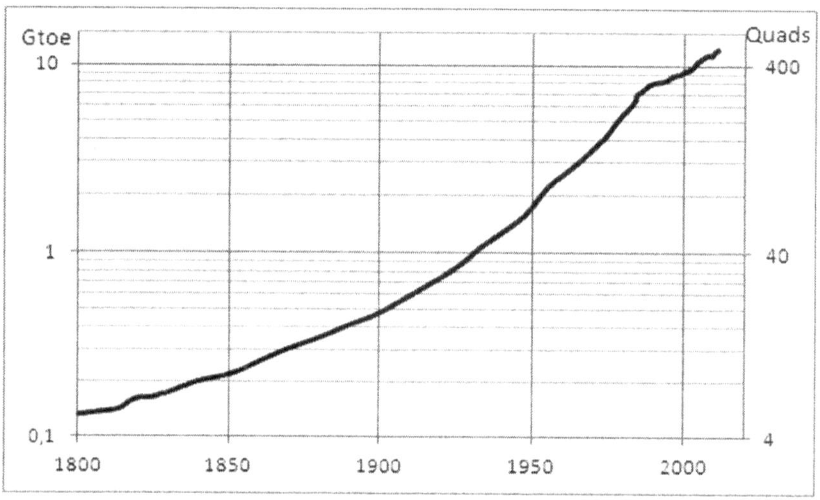

Figure 4.4 Evolution of the global energy consumption.

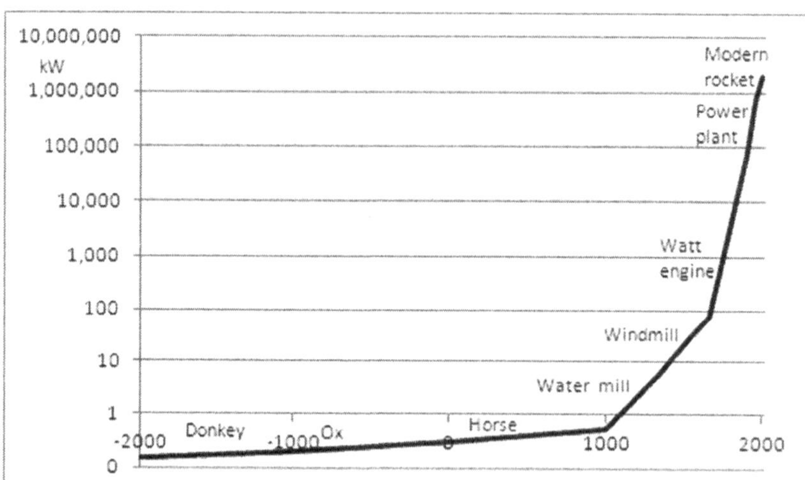

Figure 4.5 Power unit controlled by man.

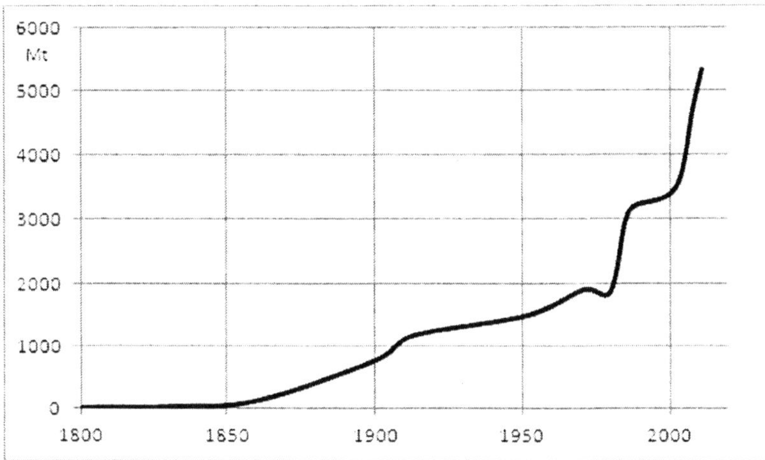

Figure 4.6 History of coal production.

4.3 COAL AND INDUSTRIAL REVOLUTION

Gradually, energy has thus become a *sine qua non* condition of economic activity. Nowadays, it is omnipresent in all developed societies, where it takes many forms and is used for many purposes. Commercial energy use has resulted in a significant increase in proportions of the average energy consumption per capita compared to the levels observed in traditional societies. This trend shows no signs of stopping because of unmet needs in developing countries.

The graph in Figure 4.6 displays the evolution of the world coal consumption between 1800 and 2010. It clearly shows that, in industrialized countries, the history of coal largely mixes up with the industrial revolution, of which it has been the main driving force.

It would however be wrong to assume that coal was the trigger of this historic event, which was made possible due to a combination of factors, such as:

- gains in agricultural productivity, which released the constraints of capital and labor;
- a technology boom (looms, use of iron, especially in agriculture, etc.);
- the low amount of capital needed during the nascent industrialization;
- an initial energy levy on English forests, coal appearing as a substitute when timber became scarce.

For further developments in this area, see the work of P. Bairoch, including the one cited in the bibliography [Bairoch, 1971].

4.4 WORK/ENERGY SUBSTITUTION

For two hundred years, energy has massively replaced labor in industrialized countries.

The basic reason was the sharp decline in the cost of energy relative to that of labor, a phenomenon that took place gradually over the last two centuries and caused energy to replace human and animal labor whenever technically and economically feasible. It should be noted that slavery could be abolished only once enough energy had been harnessed for the foundation of the economy not to be challenged by this reform.

Let us consider the case of coal.

To gain some perspective on the matter, we can estimate that in 1750, 1 kg of coal supplied 330 Wh of mechanical work and that its production cost the equivalent of about 15 minutes of human labor.

At that time, coal was already, at equal cost, two to five times more productive than human labor (taking a continuous manual labor as a base).

Two centuries later, because of technological advances (improved converter efficiencies, extraction processes, etc.), 1 kg of coal supplied about 2 kWh of mechanical work and its production required less than one minute of handwork. In two centuries, the gain in energy/manual labor substitution has thus been multiplied by a factor of 100 (Table 4.1). The upheavals that this phenomenon has effected on everyday life have been considerable.

Another illustration of possible energy/manual work substitutions is given by Figure 4.7, which shows, from 1920 to 1970, the simultaneous evolution of the

Table 4.1 Equivalence extraction time and work produced by 1 kg of coal.

	1750	*1950*
Equivalent extraction time	15 mn manual labor	<1 mn manual labor
Mechanical work produced	330 Wh	2000 Wh
Equivalent human labor	1 h	6 h
Multiplier	4	>360

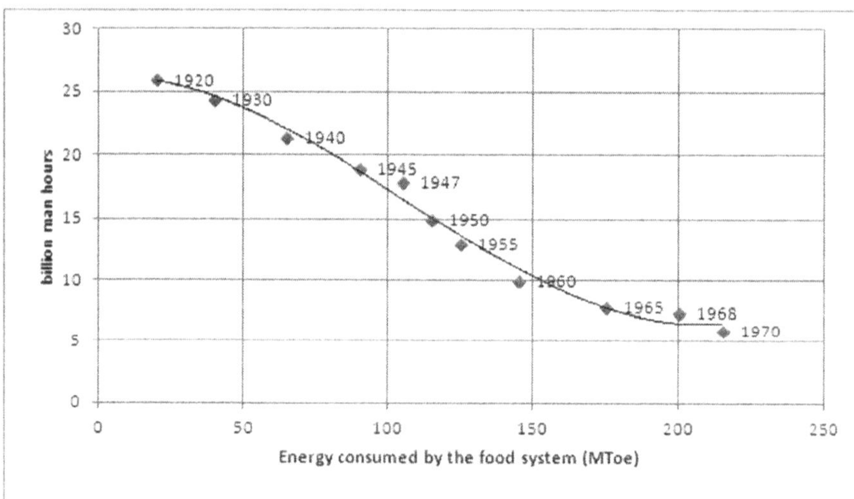

Figure 4.7 Energy-labor substitution in the agricultural sector in the United States.

number of billion hours of manual labor and the amount of energy consumed in the U.S. agricultural system. At the same time, agricultural production more or less doubled [Steinhart, 1974].

4.5 REDUCTION OF TRANSPORTATION COST

Another major consequence of the industrial revolution and the development of the steam engine was a significant decrease in the cost of transport.

Indeed, until the advent of the steam engine, transportation could be achieved by using energy of draft animals (road) or the wind (shipping).

It is considered that for a journey of 400–500 km, the cost of transporting wheat by road amounted to the value of the cargo until 1800, which means that the price doubled for this distance. With the development of railways, the fare dropped to a ratio of about 20 to 1 in a century. Meanwhile, transport safety progressed considerably (drastic reduction in robberies).

For maritime transport, the development of steamships and increased unit sizes has resulted in lower freight costs by a factor of probably more than 10 between 1820 and 1950, while the risk of piracy was significantly reduced on most seas.

The development of aviation meanwhile allowed for rapid transport over long distances, at an acceptable cost.

The impact of all this technological and economic progress on today's world is considerable. It has in particular made possible an international division of labor that was unfathomable 150 years ago, when high costs did not allow for goods transportation over long distances.

The advent of private transport has resulted in an unforeseen extension of social mobility. It has shaped our immediate environment to the point that some observers have called our era the "automotive civilization", with of course a number of well-known adverse effects.

The bicycle is otherwise the most efficient means of road transport in terms of energy consumption per kilometer. It is much more common than cars, especially in developing countries.

In conclusion, this brief history shows how the impact of energy is crucial for economic and social development.

4.6 AN ANECDOTAL ILLUSTRATION

To illustrate the low cost of energy nowadays, it is interesting to estimate the services that can cost 15 cents of euro in electricity (excluding connexion charges).

Table 4.2, which summarizes this point, speaks for itself.

4.7 ENERGY USES

Energy comes in many different forms (coal, oil, gas, electricity, non-commercial energy), but its uses are also very diverse, as shown in Table 4.3, and now involve all branches of the economy.

Table 4.2 With 0.15 euro electricity, you can:

DO	
either	one 12 place settings dishwashing → 1/2 hour to 1 hour labor saving
or	one 4/5 kg laundry load (60° with prewash) → 1 to 2 hour labor saving
USE	
15 hours	100 W filament bulb
65 hours	23 W low energy bulb (equivalent to 100 W)
15 days	vacuum cleaner (10 min/day)
either 3 weeks	disk drive (1 hour/day)
or	toaster (10 min/day)
1 month	iron (1 hour/week)
either 10 months	coffee grinder (30 seconds/day)
or	mixer (10 min/week)
2 years	hair dryer (25 min/week)
6 years	razor (10 min/day)

Table 4.3 Main uses of energy.

Sector	*Use*	*Energy*
Agriculture	mechanical energy	oil, electricity
	space heating	natural gas, oil
	fertilizers, pesticides	natural gas, oil
	drying	natural gas, oil
Industry	electrolysis	electricity
	steel	coal
	chemistry	oil
	metallurgy	natural gas, electricity
	cement	coal, oil
Residential/commercial	lighting	electricity
	cooking	natural gas, LPG, electricity
	hot water	natural gas, LPG, electricity
	space heating	natural gas, oil, coal, electricity
Transports	office	electricity
	rail	electricity
	road	oil
	air/sea	oil

Agriculture needs mechanical energy for plowing and harvesting, but also for grinding food or shelling. In temperate zones, heating of breeding facilities has appeared. Drying activities can also represent significant energy consumptions. However, one of the most demanding items corresponds to fertilizers and pesticides, which enable significant productivity gains at the cost of a high energy content, notably in oil, and of an increasingly challenged environmental impact.

Industry is one of the highest energy consumers in the economy, the share of energy costs in industrial production ranging from less than 5% to about 15%, as shown in Table 4.4. The high share of the industrial sector in the energy consumption explains why the energy intensities of countries undergoing industrialization are so high. It is probably during this phase that it is most necessary for a country to have access to cheap energy.

Electrolysis requires electricity, for example for the production of aluminum. Steel requires coke produced from coal. Many branches of chemistry depend on hydrocarbons. Metallurgy requires a high temperature heat, which can be obtained in various ways, including from gas and electricity. Cement, heavily dependent on oil until recently, can now be produced using coal.

The residential and tertiary sector is characterized by the need for low and medium temperature heat for heating, gas for cooking, and electricity for lighting and to operate household appliances and office equipment in general.

The transport sector, except for the rail market, is almost entirely dependent on oil in various forms: heavy fuel oil for maritime transport, petrol and diesel for road transport, kerosene for aircrafts.

Some kinds of energy are dedicated to specific uses, while in other cases, substitutions may sometimes be achieved. In practice, for substitutions to be possible, it is necessary that the substitution energy be available on site, which calls for the existence of a corresponding distribution network, and for other equipment used to be polyenergetic. Since the first oil shock, a major effort has been made in this direction to reduce the dependence on oil, for example by converting fuel power plants so that they can burn coal. Electricity utilities have sought to promote a number of bi-energy systems to industrials as well as individuals, to sell electricity at very attractive preferential prices during off-peak hours, while using another source of easily stored energy (usually oil) during peak hours.

The energy content provides a measure of the total amount of energy that is required to produce a given product. It makes it possible among other things, to compare different processes.

Most often, in a finished product, energy is necessary at many stages of its production, so that its energy content comes, sometimes in modest proportions, from varied forms of energy.

Table 4.4 Share of energy in production costs.

E > 10%	Steel, aluminum production Transport, glass, Ceramic, chemistry Building materials
10 > E > 5%	Non-ferrous metals Synthetic fibers Dyes, paper
5% < E	Mechanical and electrical industries Building, Public Works FPI textiles, leather Clothing, wood, etc.

For example, the energy required to produce one kilogram of white bread in Britain and to distribute it in a retail store corresponds to a total energy content equal to 20.7 million joules, or approximately 0.5 kg of oil equivalent. About 20% of this energy is used to grow wheat, 13% is used for milling flour, 47% for the actual manufacture of bread, and the remaining 20% for packaging, transport and display in the store.

Rural societies have very different consumption patterns than those of cities, often with very low commercial energy consumption. In their case, the energy requirements are provided mostly by tapping into the surrounding nature:

- solar energy for drying food;
- energy of draft animals for plowing and transport;
- firewood for cooking and heating.

The share of fuel wood in the energy balance of some countries can be considerable. This has the effect of leading to overexploitation of vegetation, especially around big cities. The firewood crisis and desertification have serious environmental ramifications. We will return to this topic in Chapter 6.

4.8 MECHANISMS OF FORMATION OF ENERGY DEMAND

To estimate the primary energy demand, consumers are grouped into homogenous modules whose unit consumption is estimated. The needs of all modules for a country or region are then added up.

4.8.1 Breaking down demand

One of the main challenges is to correctly perform the breaking down into consumer modules. With regard to economic activity, one starts from the branches retained by the national statistical institutes, which are then generally divided into large and small consumers. For households, consumers are grouped together when they share the same characteristics in terms of:

- needs to be met;
- conversion equipment;
- income level.

One thus sets up groups of quite comparable useful energy demand so that one can estimate under sufficiently rigorous conditions the changes in their consumption over time in terms of useful energy as well as final energy.

Each of these modules can be seen as the leaf of a tree that may be used to trace back the useful energy demand to the primary energy demand, the main branches of the tree representing the possibilities offered by the existing energy system and its possible evolution.

In a module, the final energy demand can be expressed by simple relationships, such as:

Productive sector:

$$D = SC \times VOL \qquad (4.1)$$

SC is the specific consumption of the product (kWh/t),
VOL is the volume of production (t).

Households:

$$D = \text{Fleet} \times UC \times UR \qquad (4.2)$$

Fleet is the fleet of existing equipment in the module,
UC is the unit consumption of the equipment in question, and
UR is the equipment utilization rate.

For the same need, the demand may thus vary depending on many factors:

- the size of the equipment fleet, which increases with household income and the development of productive activity, but can be reduced if prices increase;
- the utilization rate itself a function of the existing fleet and of the level of energy prices;
- unit or specific consumption of each apparatus, function of technological progress and in particular the efforts of rational use of energy;
- the volume of economic activity.

The aggregate demand is then calculated by adding up the demands of all the modules.

Of course, the basic laws (4.1) and (4.2) allow us, if we have the necessary data, to calculate the energy needs by type up to a high level of disaggregation. This method is the basis of the analytical models for estimating demand, such as "Medee" developed by IEPE, as discussed in section 5.4.2 on energy policy and planning.

The importance of building reliable databases on the detailed determinants of demand cannot be underestimated if one wants to seriously establish energy conservation policies. Without these data, it is impossible to make an accurate diagnosis of the situation, and thus to define appropriate policies. Moreover, the impact of these policies cannot be measured without them.

4.8.2 Determinants of aggregate demand

Now that demand formation mechanisms are explicit, it is interesting to look at the main determinants at the global level.

Given the above remarks, the final energy demand depends on:

- social needs and demographics (particularly the number of households);
- the needs of the productive sector, depending on the volume and nature of the economic activity (GDP structure).

It is indeed the number of households, more than the number of inhabitants, that determines the number of consumption units (dwellings), and thus the service equipment fleet (appliances, boilers, automobiles, etc.) and the final energy consumption.

Prices have a role in the short-term on the satisfaction of needs, and in the medium-term on inter-energy substitutions.

The primary energy demand in turn depends on the following factors:

- the final energy demand, which has been studied;
- the existing energy system, determining the modes of transformation of primary energy and efficiency;
- national energy resources, affecting the selected primary forms of energy;
- existing primary electricity, separately indicated here given the magnitude of the losses caused by the thermal generation of electricity.

4.9 MACRO-ECONOMIC IMPACTS OF ENERGY

To make energy available to multiple users, it must be produced, transformed, distributed, and sometimes stored. All of these activities, in varying degrees, impact the economy, as discussed below.

4.9.1 Energy sector

Production, transport and distribution of energy require that equipment be built, structures set up and people of various qualifications employed. Together, they form the energy sector, which accounts for a significant share of the domestic production.

For example, the energy industry represented in France, in 2007:

- 1.9% of the added value;
- 194,000 jobs, or 0.8% of the labor force;
- 2.5% of spending on research and development and 3.3% of spending in the industry.

To this we must also add indirect effects not accounted for in the energy sector, that is to say the equipment used for production, conversion and use of energy, which is very difficult to accurately estimate.

Finally, in addition to these direct and indirect effects, it is necessary to take into account the multiplier effects associated with any consumption by the agents involved in the direct and indirect activities.

4.9.2 Non-energy productive sector

As a factor of production, energy obviously has a direct impact on the rest of the productive sector, and particularly on intensive industries such as those mentioned above. Its impact is mainly related to its price and availability.

The price of energy being part of the cost of production, any increase or decrease in it is reflected in the final cost, so that the competitiveness of companies depends in part on it.

Besides its price, it is often the availability of energy that is the factor most sought after by the industry, as any interruption in its supply may stop the production or be detrimental to the production system (e.g. temperature-controlled processes, power supply of many low consumption organs that are necessary for production, etc.).

4.9.3 Impact on investment

As we saw in Chapter 2, one of the characteristics of the energy sector is the heavy investment, particularly in the electricity and gas sectors. The development of the energy infrastructure, as long as it involves substantial distribution networks, thus strains the national investment budget.

For example, the energy industry in France in 2007 represented 26% of the industry investment and 2.6% of total investments.

In the early 1980s, the World Bank estimated that the share of investments related to energy in developing countries could reach more than 40%. Updated figures for electrification are given in section 6.4.6.4.

In the Asia-Pacific Economic Cooperation (APEC), which includes 21 countries bordering the Pacific in the broadest sense, the share of investments in the energy sector in GDP between 1980 and 2001 varied from 0.5% for the United States and 1% for Japan to 2–3% for Korea and China, and up to more than 5% for Vietnam.

For these countries, the ratio of the volume of annual investment by 2020 to those made between 1981 and 2000 is estimated at 1.27 for coal, 1.56 for hydropower, 2.82 for natural gas and up to 6.62 for renewable energy such as wind or solar.

The distribution of these needs by 2030 could be as follows: 60.4% for electricity, 17.7% for the production and processing of oil and natural gas, 7.5% for their international transport, 9.2% for domestic oil and gas pipelines, and 5.1% for the production and transport of coal [APERC, 2003].

4.9.4 Impact on the balance of payments

The best-known effect of energy on the economy is no doubt its impact on the balance of payments, depending on the magnitude of the energy bill (balance of imports and exports of energy).

Indeed, countries that do not have enough domestic energy to meet their needs must supply themselves on the international market, and buy the amount needed to close the deficit. The price to pay is of course a direct function of the energy reference price, which has been that of oil for more than half a century.

For any given importing country, energy price changes h ave resulted in strong fluctuations in their energy bills, amplified or attenuated by variations in the exchange rate of the dollar, which is the currency of reference for international trade.

For example, the French oil bill rose from 9 billion 2011 constant euros in 1973 to 48 billion in 1982, falling to 12 billion in 1987 and rising again to 50 billion in 2011 (Figure 4.8). As a share of GDP, it was less than 1% in 1973, 3.8 in 1982, 1.2% in

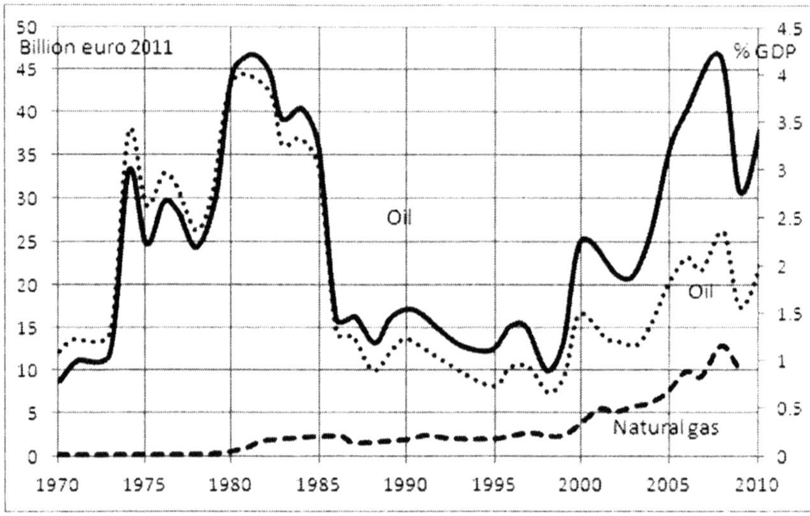

Figure 4.8 Evolution of the French oil bill.

1987 and 2.5% in 2011 (dotted curve in Figure 4.8, scale on the right). This constraint has weighed heavily on the country's economic growth, which decreased from 5.7% before 1973 to 3.3% between 1973 and 1979 and 1.2% between 1979 and 1985, rising to 2.5% in 1988 and 3.7% in 1989, before dropping to 2.2% in 2008 [CGDD, 2012].

The gas bill meanwhile increased from 0 to 12 billion euros (in 2011) between 1970 and 2011, rising almost continuously (Figure 4.8).

In tandem, energy-exporting countries, particularly those who export oil and gas, have seen their income substantially increase after each price increase. This financial inflow has allowed some countries (Algeria, Argentina, Nigeria, Mexico, etc.) to engage in significant development projects. To do so, they borrowed capital from banks that wanted to recycle the surplus related to "petrodollars", and they became all the more readily indebted that high inflation during the period concerned and prospects of high oil and gas prices made such policies economically justified.

Now that inflation is under control in industrialized countries and when oil prices were dropping, the economics of such development projects has been called into question, and these oil-exporting countries have subsequently had to face a significant risk of financial bankruptcy. Countries whose oil is a large part of their total exports are indeed facing a situation whereby the rest of their industry is penalized on the one hand due to the reallocation of resources to the oil sector and secondly because of the appreciation of the exchange rate, which makes the rest of the industry less competitive. This phenomenon is called the "Dutch disease", in reference to the situation experienced by the Netherlands a few years after the discovery of gas fields in the North Sea in the 1960s.

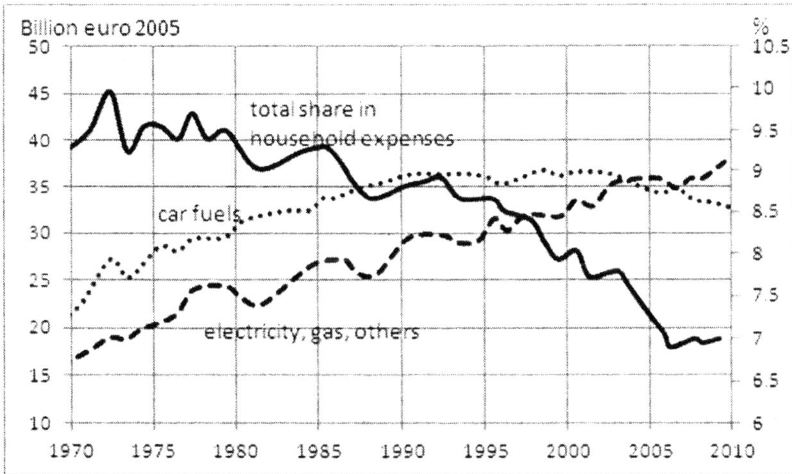

Figure 4.9 Household energy expenditure in France.

4.9.5 Impact on consumers

Energy also has an impact on the spending of consumers, for whom it accounts for a significant part of expenditure. For example, in France, the energy share in household consumption, which peaked in 1973 at about 10%, steadily decreased to about 7% in 2006 and then stabilized (Figure 4.9, solid line, right scale). To provide a perspective on this matter, these figures were higher than health spending in the 1970s. Since then, the latter increased significantly and is now close to 10% of GDP. Figure 4.9 also shows the price in 2005 euros of car fuels (dotted line), electricity and gas, and other energy household expenses in France. The former reached their peak during the 1990s, while the latter are still rising.

In developing countries, where commercial energy consumption is much lower, statistical data on this topic is scarce. We know, however, that collection of firewood may represent the work of a full-time person for a family in crisis areas.

Household consumption is of course directly influenced by the level of energy prices. Yet we must distinguish two effects:

- short-term price changes (increase or decrease) have little influence on consumption, which depends highly on the existing equipment;
- in the medium and long term, price changes affect the choice of equipment and the inter-energy substitution.

4.9.6 Impact on state revenue

Since the industrial revolution, energy has been heavily taxed by states in almost all countries. It is a significant source of revenue for national budgets.

In the United States, energy taxes levied by different states vary greatly depending on the state in question. For example, taxes linked to production (state severance

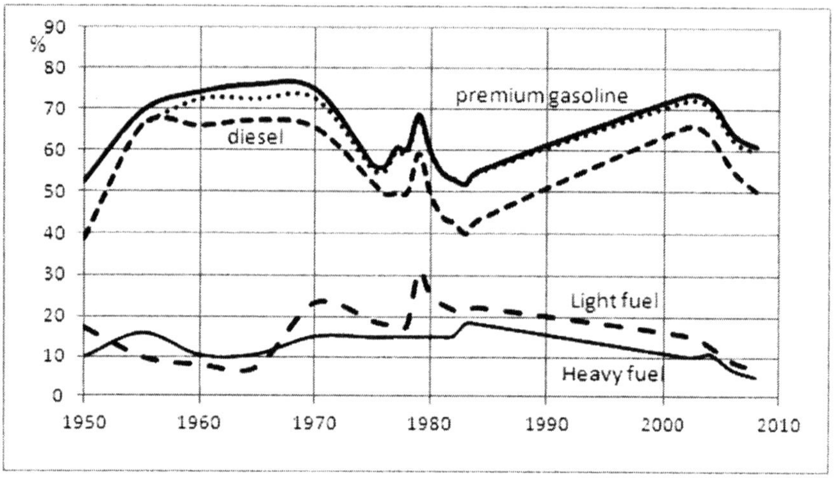

Figure 4.10 Share of domestic tax in prices of petroleum products in France.

taxes and resource leases on federal lands within their borders) represented over 65% of revenues in Alaska in 2007, and nearly 40% in Wyoming, but only 1% in Nevada [NCSL, 2012].

Moreover, federal taxes of 18 cents per gallon are levied on gasoline and diesel fuel used in road transport.

In France, these revenues are essentially twofold:

- the domestic tax on petroleum products (TIPP for *taxe intérieure sur les produits pétroliers* in French), which represents between 85% and 90% of total customs revenue. In 1986, its share of state resources was 9.1%, and in 2011 5.5%;
- the value added tax (VAT) on other forms of energy, including electricity.

It should be noted that the TIPP has been widely used by governments to balance out fluctuations in international prices. It is one of the adjustment variables of the pricing of energy, which in turn is one of the components of energy policy as discussed in Chapter 5.

In such conditions, the consumer energy price changed little in France compared to the international market, the pricing policy being used as a buffer.

This was made possible because the taxes in some cases constitute a very important share of the sale price (60–75% in the case of premium gasoline at certain times, for example), as shown in the graph in Figure 4.10.

REFERENCES

APERC, Energy investment outlook for the APEC region, Tokyo, 2003.

Bairoch P., Le Tiers Monde dans l'Impasse, Idées, Gallimard, 1971.

CGDD, Chiffres clés de l'énergie, Commissariat Général au Développement Durable, Paris, 2012.

Cook E., The flow of energy in industrial society, Scientific American, 224(3):136, 1971.
Fericelli J., Lesourd J. B., Énergie: Modélisation et Économétrie, Economica, Paris, 1985.
Galbraith J. K., Salinger N., Tout savoir ou presque sur l'économie, Editions du Seuil, 1978.
Steinhart J., Steinhart C.,"Energy Use in the U.S. Food System", Science 184:307–316, 1974.
NCSL, State Energy Revenues Update, National Conference of State Legislatures, [Online] Available from: http://www.ncsl.org/issues-research/budget/state-energy-revenues-update.aspx/[Accessed 12/26/2012]

FURTHER READING

Chevalier J.-M., Barbet Ph., Benzoni L., Économie de l'Énergie, Presses de la Fondation nationale des Sciences Politiques et Dalloz, Paris, 1986.
de Jouvenel B., La Civilisation de Puissance, P., Fayard, 1976.
O'Toole James, Energy and Social Change, The MIT Press, Cambridge, Mass., 1976.
Percebois J., Économie de l'Énergie, Economica, 1989.
Radanne P., Puiseux L., L'Énergie dans l'Économie, Syros, Paris, 1989.

Chapter 5

Energy policy

5.1 PRESENTATION

In the previous chapters, we have shown that:

- energy has been and remains one of the main development factors. Even if specific consumptions are reduced today compared to the past, the economic activity and meeting socio-economic needs require a lot of energy;
- the energy sector is rigid, implementation delays of projects being very long (5 to 10 years minimum);
- the very high amount of needed investment and energy distribution constraints, especially in the case of electricity and gas, require in many cases an oligopolistic or monopolistic structure.

All these factors, to which were added threats of energy disruption during the last four decades and the impact on the balance of payments of sudden price changes, led governments to devote much more attention to energy policy than previously in all countries, be they exporters or importers of energy.

Focused on energy supply during the years that followed the first oil shocks, energy policies gradually began to deal with a second, complementary component: demand, which concerns not only a small number of operators but also the multitude of users. Leading to results beyond the initial expectations, these new policies of energy conservation are now fully accepted.

5.2 LIMITATIONS OF THE ENERGY POLICY

For the reasons outlined above, in any country in the world, the State cannot ignore the energy issue. In producing countries and energy exporters, the government tends to retain control of the exploitation of the subsoil, which is an issue of national sovereignty. In countries that are heavily dependent on external energy, the State intervenes to ensure the security of supply.

However, the development of the energy sector can often be entrusted to private or public operators autonomous enough to work effectively.

One of the challenges for defining an energy policy lies precisely here: how to reconcile, and to what extent, the concerns of sovereignty and national security with those of cost and efficiency.

Defining an energy policy, for these reasons, is a dialectical process at the interface of the political and economic spheres.

An energy-exporting country faces two major issues:

- At what rate should it export its energy? Resources being limited, with reserves on production ratios R/P ranging from almost 95 years (Iran and Kuwait) to less than 10 years (North Sea), is it better to use the energy assets for accelerating today's development, or to keep them for future years, either for domestic consumption or for foreign exchange earnings?
- What should be the share of exported production compared to that of domestic consumption and what prices should be chosen internally?

For energy-importing countries, energy policy should generally have three main objectives:

- diversify imports, both in terms of energy sources and of geographical origin, so as to minimize political risks;
- reduce the energy content of GDP, playing on the one hand on the structure of production and on the other hand on the energy content of the products;
- increase the domestic supply of energy.

The main difficulty of energy policy is that on the one hand the risks are many, some exogenous variables such as the international price of energy being likely to fluctuate rapidly, and on the other hand the inertia of the energy sector is enormous. To reach the optimal balance between the energy system and an uncertain future under these conditions may appear challenging.

The role of an energy policy is therefore to understand as accurately as possible the existing uncertainties, in order to ensure the relevance and the robustness of the choices made in the medium and long terms.

Making a choice is made even more complicated by the fact that energy operators are often competitors, and because of the possible inter-play of energy substitutions, so that the decision-making process is subject to much controversy, requiring arbitrations in terms of market shares or investment authorizations.

The energy planning process should focus on strategies to make often-conflicting energy producers and distributors align with one another to meet the needs of users, who, widely distributed and diversified, are not able to defend their interests as well as the former, in light of societal choices.

At one end of the chain, there is indeed a small number of powerful operators, with expertise and considerable analytical means, often exceeding those of state agencies responsible for their control, and with the capacity to lobby in political pressure groups. At the other end, there are a multitude of users, hardly able to make their voices heard because of their dispersion and the complexity of the issues at hand. These same users also have a key role to play with regard to the Rational Use of Energy (RUE) and the use of renewable energy.

With such diverse interests at stake, the energy policy strives to find the best overall coherence vis-à-vis both the sector plans of the various operators and the rest of the economy and society.

In the interest of sound management, considering the time of implementation and investments, operators have to develop long-term sector policies, taking into account their own technical constraints and the evolution of a number of variables that they are forced to consider exogenous in their analyses:

- demand;
- prices;
- environmental constraints.

As they are unable, at their level, to take stock of the entire economic system or independently fix the value of exogenous variables in accordance with overall policy orientations, operators have historically tended to overestimate their role. Even when a consensus emerged among them, it appeared that the overall consistency was not necessarily assured.

One of the main interests of energy planning exercises is that they require all energy sector players to work together and, in so-doing, to explain their hypotheses and their forecast methods.

The adoption of a common language helps clarify the contradictions of the system and thus offers the possibility of better-informed central arbitration. But more importantly, it allows actors to adopt consensual positions regarding some fundamental choices. This exercise greatly limits the risk of technocratic decisions, promoting the democratic participation of all actors.

Consistency vis-à-vis the overall economy requires that energy-economy interactions be properly taken into account:

- the investment program depends on the chosen energy scenario, including the share of goods manufactured in the country and the share of those imported. These investments are a financing constraint, which is closely linked to the external debt. They impact savings, and for the investments made in the country, employment and production;
- the balance of payments is directly influenced by the energy strategy: energy import bills and cost of funds borrowed from abroad;
- energy pricing has an effect on the purchasing power of households, which has a direct impact on their savings and consumption. It also determines in part the costs of production, especially in intensive sectors. Therefore, it affects directly the competitiveness of enterprises, that is to say the GDP, through the economic activity, and external constraints through imports and exports, which depend on the overall international environment. It also has an impact on state revenues.

5.3 MAIN OBJECTIVES OF THE ENERGY POLICY

As stated above, the energy policy of energy-importing countries generally pursues three main objectives:

- to diversify imports;
- to reduce the energy content of GDP;
- to increase the domestic supply of energy.

Note that the latter two concerns are also generally shared by producing countries, who additionally need to manage their assets, i.e.:

- set the desirable pace of exports, and
- arbitrate between exports and domestic consumption.

Diversifying imports is possible by substituting a predominant source of energy (usually oil) with other sources that are deemed safer and equivalent in terms of use. It is certain that at present oil is the most versatile and flexible energy source, but there is a considerable flexibility in term of substitutions. The development of transport over long distances and distribution of natural gas gives the latter fuel attractive prospects. Political diversification is also possible because of the existing sources of supply throughout different parts of the world, such as, for Europe, in the Mediterranean Basin and the Commonwealth of Independent States (CIS).

Coal has the advantage of large reserves, which hold at bay the specter of scarcity. However, pollution constraints and the cost of the needed infrastructure do presently limit its use to large consumers. Used primarily for the production of electricity, its supply is of little political risk in the OECD.

The range of accessible types of energy therefore remains sufficiently diversified. The extent of the consequences of the oil shocks was for many the result of the dominance of oil in imports.

Reducing the energy content of GDP is the second option. Concerns about energy conservation, minimal when energy prices were very low, have been coming to the fore since 1975. Between 1986 and 2006, due to the oil countershock, they somewhat lost their relevance again. However, the renewed interest in environmental issues gave them a second wind. As such, the concerns returned to the fore when the price of oil increased.

Reducing the energy content of GDP is possible through structural changes in the productive sectors (reduction of the share of industry, growth of services), waste reduction (no more heating of unoccupied facilities during weekends), and especially the adaptation and renewal of energy processing and consuming equipments, as well as industrial processes.

Changes in behavior are also important: the notion of comfort, which is highly subjective, has evolved considerably in recent decades, and it is now again considered normal to wear a sweater in winter, etc.

Energy conservation correspond to fields where gains can be made at very low or zero cost, but also to some that require significant investments related to the modification of the equipment used. In the latter case, the rate of achievement of savings is the rate of fleet renewal. This is for example the case for insulation of old buildings or the introduction of low consumption vehicles. Tax incentives or government subsidies may be necessary to stimulate investment.

For instance, the *Grenelle de l'Environnement* in France has implemented a number of incentives such as premiums for the destruction of old vehicles or bonuses and penalties based on cars' CO_2 emissions.

In the industrial sector, where energy conservation concerns are older than most people think, the energy cost is often a small part of the added value, well below wages in all cases (except for air transport).

In the six industries that consume most energy (steel, chemicals, cement and building materials, glass, paper, aluminum), where the cost of energy accounts for about 15% of the added value, the motivation is strong. By the 1970s, all easily achievable energy savings had been made. Subsequently, these industries had to come up with more energy-efficient manufacturing processes, using more efficient technologies.

The increase in national energy supply is the third possible path. As the review of the comparative costs of different energy sources showed, the potential of the national energy supply is all the greater as energy is expensive. High-energy reference prices not only promote energy savings for new investments (capital/energy substitution), but they also enhance the profitability of alternative energy sources (nuclear energy, new energy, unconventional fuels, etc.).

Therefore, this possibility increases in the same direction as the constraint, but with significant delays.

Note that both energy savings and the development of national energy generally have a positive impact on employment. In this sense, the energy conservation policies have macroeconomic benefits *a priori* larger than those focusing on energy imports. This remains true as long as they do not lead to a distortion in prices that penalizes the competitiveness of other productive sectors and as a levy on their available savings does not limit investment opportunities in other sectors.

The financial constraint should be considered holistically, and allocation of available resources should maximize their impact on the overall economy.

Energy is not and has never been an end in itself. It is a production factor and a means to better the quality of life, which is quite widely dependent on the level of development, at least up to a certain threshold.

5.4 ENERGY POLICY

Energy policy must nowadays consider four broad classes of problems:

- the coexistence in the energy sector of a small number of operators responsible for most of the supply policy and a multitude of users, points of application of the energy management policy;
- the presence of highly uncertain exogenous variables, such as international energy prices and the global economy;
- the competition between the different energy sectors, through the set of possible substitutions for the same end use, the selection criteria taking into account not only the economics of projects, but also other aspects such as independence, vulnerability, impact on employment and on the environment;
- the overall economic and social impacts of the energy policy.

5.4.1 Planning process

Under these conditions, the planning process is a forecasting exercise aimed at quantifying as much as possible the economic and social impacts of possible energy policies in order to guide as well as possible decision makers in their choices.

Faced with the increasing complexity of energy-related problems, a set of planning methods has emerged, assisted by the use of computer models, some of which are presented below.

The purpose of energy planning is to optimize the evolution of the energy system based on a set of economic, social and political issues, while defining the steps to bring about this change.

To do this, energy planning is under very strong constraints. In particular, given the long lead times for implementation of energy projects, which are the rule in exploration, production, distribution and RUE, energy planning can only be considered in the medium and long terms, while the international environment fluctuates strongly in the short term and a forecast error may, as we have seen, have significant macroeconomic impacts.

For this reason, after a period of time when energy planning followed a linear path centered on the supply, calculated for a hypothetical demand that was assumed to be exogenous, a new era has now begun over the last few years, which emphasizes reasoning in terms of systems analysis, an iterative process involving several nested steps.

Depending on authors and situations, different variants are possible. In the context of this introductory book, we will retain a somewhat schematic approach involving three basic steps (see Figure 5.1), referring readers interested in further developments to the specialized literature.

1 Assessment of potential. A prerequisite for optimization is to know as well as possible the starting point, to assess the initial state of the system. This is the purpose of the diagnosis. In terms of demand, the analysis of past trends allows one, within certain limits, to anticipate the future, even taking into account behavioral changes in the productive structure or unit consumption. On the supply side, the resources of the country, the state of the national energy system, and the level of technological development determine the main options for the future.

   By integrating demographic, social and economic trends as well as probable changes in prices, through contrasting scenarios or by making the variables endogenous, one can get an idea of the paths available for the entire energy system.

2 The selection of the main options is the next step. Among all the possible developments, a choice must be made. To do this, one needs to integrate the non-economic criteria that we cited as well as the impact of the considered solutions on the overall economy.

3 Policy implementation is then possible: detailed analysis of projects, investment planning, development of energy conservation policy, pricing and definition of institutional capacity to implement.

It is clear that the third stage impacts the first through prices and the second the first through the consequences of the scenarios on the overall economy.

This process is necessarily iterative. Given the complexity of the relationships between the three steps, it is greatly facilitated by the use of computer models.

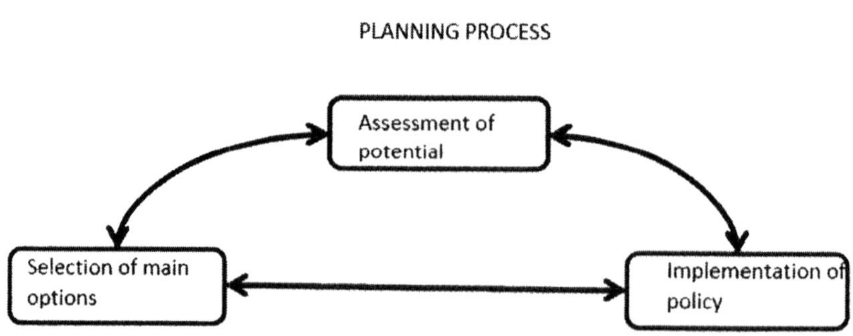

Figure 5.1 Planning process.

5.4.2 Energy models

To understand the behavior and evolution of complex systems, models are a valuable tool.

Energy models can be classified into two broad categories:

- supply models;
- demand models.

5.4.2.1 *Supply models*

Supply models, which are in their vast majority energy producer tools, most generally aim at:

- optimizing the management of the existing fleet: management of distribution networks (electricity, gas, district heating), production management (electricity and refining);
- optimizing investments (electricity, refining, gas, etc.);
- calculating the primary energy demand;
- calculating the level of pollution caused by extraction and processing.

The limits of supply models are mainly threefold:

- a total exogeneization of the demand, while in reality it depends on the supply, at least partially, even if only through the price;
- an imbalance between the model complexity (reflecting the complexity of the sector's technical management) and the great uncertainty of assumptions regarding both energy prices and the structure and level of the demand;
- finally, modeling is a subjective art, and supply models frequently reflect the assumptions of their designers.

For example, we will detail the operation of a highly simplified model of electricity supply.

The basic data is called the power curve, which is presented as a series of electrical power amounts required by the network (see Section 2.5.2). Being rather inconvenient to use, electricians are accustomed to transform it by summing demand hours corresponding to a given power.

This leads to the monotonic load curve or "demand monotone" (Figure 5.2), where the abscissa gives the number of hours h during which power demand was greater than the value shown on the ordinate (it is a cumulative frequency curve). On this curve are placed the equipment depending on their cost and proportional fixed cost [Gomez-Exposito et al., 2008]. The aim is for the total cost (fixed cost plus the proportional one multiplied by the number of operating hours) to be minimal.

We have shown the investment and operating costs of some power generation equipment in Table 2.4 reproduced here as Table 5.1.

It clearly shows that the most expensive equipment investments are those whose running cost is lowest. The basic principle is to operate the low-proportional-cost equipments during off-peak hours and keep the low-investment and high-proportional-cost ones for the peak hours.

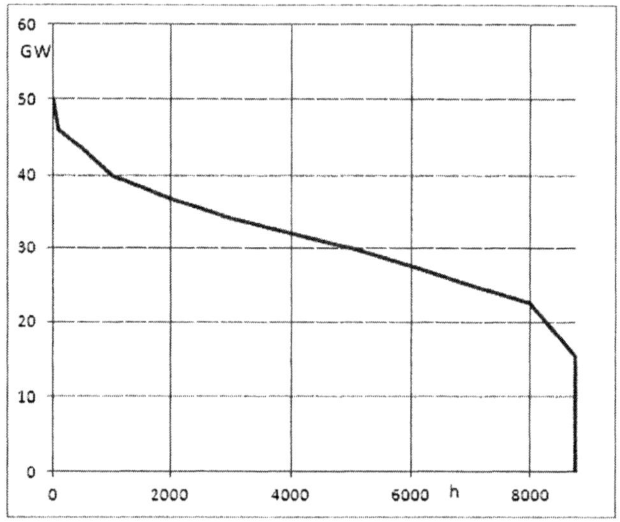

Figure 5.2 Monotonic load curve.

Table 5.1 Production costs of electricity.

	Unit size MW	*Cost of invest. \$/kW*	*Fixed costs \$/kW*	*Operating costs \$/MWh*
Gas turbine	150	730	64	140
Combined cycle	480	1070	95	84
Pulverized coal	750	2130	180	42
Nuclear	1400	4100	330	

Fixed costs C_f are the sum of discounted investment cost and equipment maintenance costs. The proportional cost C_p is essentially the cost of fuel.

Graphically, one classifies the various means of production based on their total annual operating cost C_t, related to the number of operating hours N_h by equation (5.1):

$$C_t = C_f + N_h \times C_p \tag{5.1}$$

Intercepts of cost lines define the optimal number of hours of operation of each means of production (Figure 5.3). Thus, with the figures in Table 5.1, it is more economical to use gas turbines below 500 h, combined cycles between 2000 h and 500 h, coal power plants between 2000 h and 5500 h, and beyond, nuclear plants.

One then reports the optimal number of hours of operation on the monotone, which sets the required power levels (Figure 5.4).

The production fleet is then dimensioned. In this schematic example, there should be 28 GW of nuclear power plants, 7 GW of coal-fired power plants, 6 GW of combined cycles and 8 GW of gas turbines.

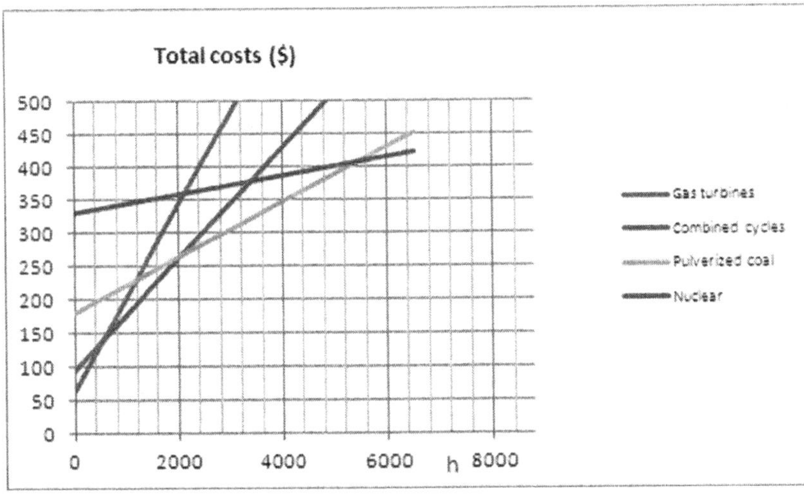

Figure 5.3 Total costs.

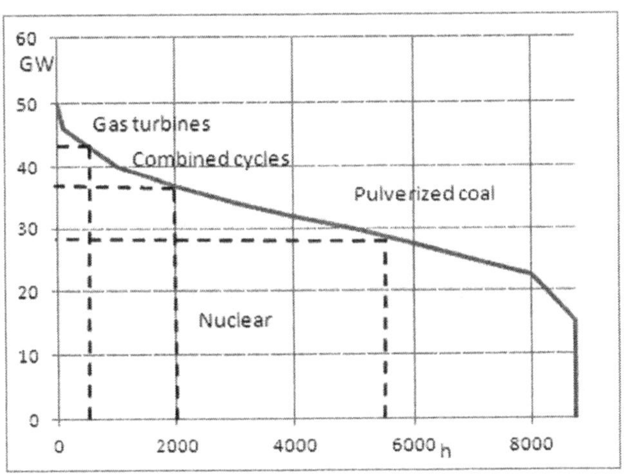

Figure 5.4 Placement of the means of production.

This model, based on the monotone of the annual demand, is very simplistic and does not take into account pumping or predictive maintenance.

In reality, the process is much more complex: to select investment strategies, one does not rely on the annual demand, but rather on a series of 52 weekly power demand monotones.

For each of these monotones the output is calculated with fixed equipment, which necessitates the placing of pumps and turbines, whose 0.7 efficiency allows the substitution of electricity at peak hours by electricity at off-peak hours, when it is financially relevant (pumping when there is a surplus of power in nuclear power plants and when retention ponds are not full, operating turbines during the peak hours if the basins are not empty).

One then calculates the weekly cost and, overall, tries to optimize a cost function such as:

$$C = \Sigma_i(C_{P_i} X_i) + \Sigma_i g(X_i) \tag{5.2}$$

To take into account the various uncertainties (fleet availability, weather conditions), the summation is extended to a large number of cases, weighted by the probability of occurrence:

$$C = \Sigma_i(C_{P_i} X_i) + \Sigma_i g(X_i, \text{hazards}) \times P\ (\text{hazards}) \tag{5.3}$$

5.4.2.2 Demand models

Demand models are used to estimate the future energy consumption, overall or by source. Historically, they were essentially based on deterministic methods keyed to compliance with the laws of the past. Since the oil shock and counter-shocks, more accurate methods have been developed to try and take into account the observed decoupling between economic production growth and that of energy consumption.

Very generally, the demand models primarily aim at:

- defining the structure and the level of demand;
- calculating the non-energy factors and goods demands (mainly energy using equipments);
- calculating the level of pollution in the final consumption.

Depending on the model, the modeled behavior and choice of exogenous variables are very diverse.

Very roughly, we can however distinguish two broad classes of demand models:

- analytical models;
- econometric models.

These two types of models are complementary rather than competing, as each provides a partial view: the analytical approach provides a framework for precise causal analysis, while econometrics explains the evolution of each of the used parameters (stocks of materials, activity, etc.), based on variables such as prices or macroeconomic indicators (growth, employment, incentive policy, etc.).

In analytical models, one represents the entire energy chain by a graph linking the useful energy to the primary energy, disaggregating the consumption chain as finely as possible given the available data.

The goal is to connect energy to an activity indicator by a law of the type (5.4) for each consumption module i:

$$E_i = EI_i \times A_i \tag{5.4}$$

Energy E_i of module i is equal to the product of energy intensity EI_i by activity A_i.

The model then performs the aggregation by summing over all modules i (as explained in section 4.8 on the demand formation).

Its value lies in the fact that it allows for an easy manipulation of all these indicators within a framework that should reflect as closely as possible the total consumption.

It is important, and this is undoubtedly one of the most delicate points, to ensure that all data are consistent with one another and that the outcome of the model is itself consistent with the explicit or implicit macroeconomic assumptions.

This approach has the great merit of providing a unique tool for the development of energy conservation policies. To some extent, it can be validated experimentally, as the technical coefficients used are sometimes measurable.

The Médée model developed by IEPE Grenoble and marketed by Enerdata is an illustration of this method.

Limitations of analytical models are mainly threefold:

- the risk of inconsistent data, which is greater when the model user is less experienced and his/her knowledge of the model content is lower. In expert hands, this risk may disappear;
- the difficulty of taking into account the dynamic changes (technical coefficients, equipment fleet, etc.), which must be made exogenous;
- they only partially take price effects into account.

Moreover, some analytical models simply simulate scenarios, while others seek to optimize the studied system [Zagame, 2008].

Econometric models in turn connect the energy consumption to activity, income, relative prices and sometimes technological indicators.

The purpose of these models is to represent the role of energy in the economy in its economic rather than technological dimensions.

The emphasis is more strongly put on the formal representation of the behavior of substitution between factors of production (capital, labor, energy, raw materials), and the price effect may be taken into account in much more detail. However, their deterministic nature form an obstacle to their use in a quickly changing international economic environment.

5.5 PRICING ELEMENTS

Pricing energy is a problem that is more the responsibility of economists and financiers than technicians. In this chapter, we do not pretend to exhaustively cover the issue. We only wish to provide the basic elements necessary for further reflection. The price issue is indeed very complex: it includes multiple aspects and affects the overall energy policy. As one of the most direct tools available to governments, the pricing policy features all the ambiguities of the energy policy, which is, as we saw, an eminently dialectical process.

Pricing requests that at least three basic considerations be taken into account:

- the economic efficiency of prices;
- their impact in terms of social equity;
- the financial viability of the energy sector.

The effectiveness criterion is here to ensure that the microeconomic decisions of consumers have an overall result as close as possible to the macroeconomic optimum, whereas the equity criterion aims at ensuring that the poorest can access enough energy at an acceptable cost for their basic needs. The criterion of financial viability of energy operators assumes that they can get sufficient revenue from their activity to cover their costs and prepare for the future.

There may be contradictions between these objectives, especially between the social equity and economic efficiency or financial viability criteria. The choices vary a lot depending on the country, resulting in varied economic and social impacts. As an illustration, Table 5.2 gives some values of energy prices for end users in Europe in 2007 [Eurostat, 2008].

Significant divergences in tariff policy can thus be illustrated.

In addition to the three fundamental considerations mentioned above, two other factors must also be taken into account:

- the price level determines the threshold of profitability of alternative energies and energy savings: high energy prices foster energy efficiency and the development of national energy potential. In contrast, low prices can boost the competitiveness of energy-intensive industries;
- the energy sector, for various reasons, is one of the main sources of tax revenue in many countries, in the form of rent taxes in the producing countries and indirect taxes in consuming countries.

5.5.1 Economist pricing approach

The objective of the economist pricing approach is to determine the pricing based on the economic theory: the price structure thus depends on the cost of procurement, production, storage and distribution.

This approach has the merit of consistency with the overall economic policy. One can move away from its findings but may so introduce various price distortions sometimes difficult to correct, which are important to identify.

The basic scheme is shown in Figure 5.5. Prices consist of three distinct elements in this approach:

- international prices, an exogenous variable;
- technical costs, broken down into "supply" costs and "user" costs;
- the economic policy, in particular taxes.

The price sought for energy should reflect this structure.

In practice, there are two possibilities to set the prices of various energies: the average cost pricing and the marginal cost pricing.

Table 5.2 Energy price in Europe.

	France	*Germany*	*U.K.*	*Spain*	*Denmark*
Gasoline 95 (€/l)	1.2	1.22	1.3	0.9	1.2
Diesel (€/l)	1	1.05	1.4	0.9	1
Electricity (€/kWh)	0.13	0.19	0.14	0.12	0.26
Natural gas (€/GJ)	14	19	12	14	31

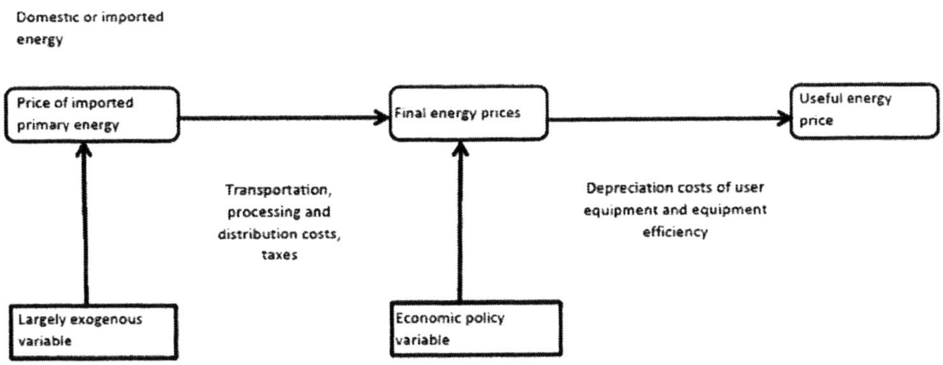

Figure 5.5 Economist pricing approach.

5.5.1.1 *Average cost pricing*

On behalf of the equality of all citizens before the public service, a uniform pricing is established for the whole country anywhere and at any time.

If the egalitarian principle of the single price is quite attractive on a philosophical level, it has the disadvantage, however, of not reflecting the actual cost that each user inflicts on the national community.

Examples of this type of pricing are multiple. Their benefit is well known, but they can in some situations lead to economic aberrations. Thus, for example in France, electricity is sold at a price much lower than its production cost in some overseas departments.

5.5.1.2 *Marginal cost pricing*

The principle here is to make the user pay a price that reflects the cost actually incurred by the community because of its presence on the market.

The price is not set according to whim, but reflects the cost related to the client's specific energy demand.

A system where prices are modulated between different classes of users is derived from this principle: e.g. pricing depending on the voltage level, on the time of day (peak, off-peak) or the season (summer or winter), etc.

Such a system has the advantage of not introducing distortions in the cost structure and thus makes it easier to reach a financial balance of the operators, the third major consideration set out at the beginning of this section.

However, it does not include any element of social equity nor a redistributive concern. It is therefore often necessary to amend it to correct these deficiencies. For example, the French electricity pricing is derived from both social equity and marginal cost concerns: rates are modulated based on the average demand of different categories of users, but the price is the same for all users in the same category.

5.5.2 Political approach to pricing

An opposite approach to the economist one is to deliberately ignore the cost structure and focus only on purposes. Once the objectives are defined, prices are set so as to guide policy choices of the users towards these goals. The concern of economic orthodoxy is secondary to concerns of national independence, social redistribution, industrialization, land, etc.

Either through subsidies for domestic production or through high import taxes, it is possible for the government to introduce distortions that make a type of national energy competitive even though it could not be so without these "boosts".

However, it should be noted that these distortions always have a cost, which must be borne by the taxpayer or by the user:

- in the case of subsidies, the taxpayer bears the cost of providing a domestic supply at a price lower than its production costs. The user gets energy at market prices;
- in the case of import taxes, tax revenues increase, thus relieving the taxpayer. The user, however, is penalized by energy prices that are higher than the market price.

In both cases, the result is that net imports fall, which is the goal, but the cost of this measure is not supported by the same social group.

5.5.3 Practical difficulties of pricing

In practice, tariff policy is a compromise between the three contrasting attitudes that we have studied:

- average cost economic approach;
- marginal cost economic approach;
- political approach.

Besides, the implementation of a tariff policy faces several difficulties:

- uncertainty regarding international prices. The international price of energy, one of the major exogenous variables, is likely, as the last four decades have shown, to fluctuate dramatically. By the interplay of taxation, governments can mitigate the short-term changes, downward or upward, of the energy price, and thus smoothen to some extent the effects of shocks on their savings. However, their flexibility remains low, if they do not want to create distortions in the medium and long terms, often affecting the equilibrium of the whole economy;
- the need to consider the cost of the equipment. As we saw when we examined the mechanisms of demand formation, what is important is to meet the need, and this need requires energy-consuming equipment in order to be satisfied. Any policy encouraging inter-energy substitution through prices must take into account this factor. Neglecting it may divert an incentive policy from its purpose and benefit to the more affluent social strata, able to finance the necessary equipment, rather than the lowest incomes;
- the existence of external effects difficult to estimate beforehand. The energy sector permeates the whole economy. As such, it strongly interacts with it. We have seen that many planning models suffer from limitations regarding this interaction.

To consider most macroeconomic variables as exogenous is a trick that is justified in part by the economy of complexity it provides, and also by the lack of precision of coupling models.

However, any energy policy has important effects on the economy (debt, unemployment, levy on savings, revenue, pollution, etc.) that can rarely be accurately estimated.

It is therefore highly desirable to be cautious in the evolution of pricing policies.

It is often better to adapt the system in place by successive modifications rather than seeking to establish a radically new pricing system from scratch. Iterative steps make corrections easier if unexpected negative externalities appear.

5.6 INSTITUTIONAL ASPECTS

5.6.1 Need for joint action levels on supply and demand

After the oil shocks, it appeared that an action only on supply, that had characterized government interventions until the early 1970s, was not enough. It was certainly necessary to continue to look for alternative energy solutions and better diversify sources of supply, but it was also required to reduce demand without jeopardizing economic growth.

The energy problem appears as a multifaceted problem requiring multiple and coordinated solutions, conventional sector responses limited to the supply side having become insufficient because of inadequate regulatory mechanisms internal to the energy system.

As the energy problem cannot be solved at the technical level or by a simple rearrangement of institutions, it was necessary to reconsider the relationship between the user and the resource. This required policy choices more complex than before, affecting society as a whole.

This finding raises several comments:

- solving its energy problem leads a society to ask fundamental questions. It presupposes an effective political mobilization at the highest level since only a strong and resolute political will can enable the implementation of necessary reforms;
- governments must also have the effective capacity to enforce these reforms. This is all the more necessary given that significant results will occur only after a fairly long time, while internal controversies arise immediately. Debates on nuclear energy in many countries bear witness to this;
- structural adjustment policies are already difficult to achieve during periods of strong growth, but even more so in times of crisis, because they then translate into reallocating scarce resources and challenging the established order and the privileges of certain social categories;
- technical answers are many, but compete with one another. Except in extraordinary circumstances, one cannot expect a spontaneous consensus after an analysis and planning exercise, however complex and detailed it may be. The technical and financial decisions sometimes lead to violent disputes between conflicting interests, which have to be mediated by the political authority;

- political reforms go hand in hand with institutional reforms. Decisions have a value only if the institutions are able to implement them, and sound institutions are powerless, or even harmful, if the law thwarts their effectiveness. The need to break with traditional sector approaches only becomes a reality if it is accompanied by a reform of the entire energy sector.

In practice, the solution to the energy problem thus requires a combination of carefully balanced sector policies and a policy of demand management in all sectors of the economy.

The success of such a project requires a reorientation of the energy policy and a redefinition of the traditional roles of various sector operators.

Low-energy research and development strategies allowed many countries to acquire additional degrees of freedom and to avoid macroeconomic deadlocks to which purely supply-oriented policies without any action at the demand level would have led.

Energy planning has a vital role to play in this context: it can enable a country to learn how to undergo an energy transition, and it helps guide the policy choices.

This dual approach is also appropriate for traditional fuels. As we will indeed see in section 6.4, supply-side policies are insufficient to resolve the firewood crisis, for they are impossible to implement at the required level. One of the main problems is indeed the difficulty to generate active participation of the population in time. Traditionally, wood was a free community resource, whose opportunity cost was low. In areas of scarcity, this is no longer the case (the World Bank estimates at 250–300 man-days the annual work of collecting wood for a family of five people in some regions of Africa).

However, due to the lack of monetary incentive in the non-monetized subsistence sector, efforts to redress the situation rarely succeed before irreversible ecological degradation occurs. Again, a sector policy is doomed to fail, and public action is necessary, if possible supported by Intergovernmental Organizations (IGOs) and Non-Governmental Organizations (NGOs). More effective methods remain to be designed for substantial progress to be made in these areas.

In countries with a deficit, firewood should be one of the major concerns of the national energy policy. Due to the lack of sufficient political mobilization on this issue and of institutional reforms, the problem will worsen, with a negative economic growth in rural areas as a corollary, as well as an increased rural exodus, exacerbated disparities between town and country and the whole procession of explosive political risks generated by such a situation.

5.6.2 Organization of national energy systems

5.6.2.1 Central administration

The problem of the organization of a national energy system first arises at the level of the State: should it have a single administration in charge of energy?

If we look at what is happening in the world, there is a powerful trend in this direction, but it can take various forms.

In many countries a Ministry of Energy has been created, often referred to as the Ministry of Energy and Mines. Other have opted for the establishment of an autonomous Energy Agency, more independent from the political authorities, and also more decentralized. Some examples are given in Chapter 7, where eleven country case studies are presented.

5.6.2.2 *The energy operators*

The organization of the energy system concerns both the governmental supervisor and the energy operators themselves. At this level, the organizational problems are much more numerous and complex.

One can try, in a kind of methodological approach, to group them around a number of alternatives:

1 The first problem is that of the respective roles of national operators and foreign operators. It arises especially in the field of hydrocarbons, due to the massive concentration of the world industry around a small number of very powerful multinational corporations. It also arises in other sectors, such as coal mining.

2 The second question, which in practice often overlaps with the previous one, although it is different, is: should we foster the establishment or development of public or private operators?

   The problem consists here of two separate issues:

   - Should the State become an entrepreneur in case there is a market failure?
   - Should the State nationalize private operators, when they exist?

3 The third dimension of the organization of a national energy sector is that of monopoly or pluralism: should one, for each type of energy, promote the establishment of a single company, or otherwise seek the plurality of operators? This question may arise differently in each case, at the stage of production, transport and distribution of energy.

   The creation of a monopoly may answer a concern for rationality, simplicity, safety (nuclear) or economies of scale. Pluralism can, in turn, be sought for very different reasons: to promote competition, of course, but also to enhance the security of supply of a country by fostering the emergence of national public poles facing foreign operators, or to organize the country's decentralization by developing a plurality of regional operators who share the market based on geographical localization.

   It seems that since the 1980s, many countries or groups of countries seem to consider that the best way to promote both the interests of consumers and the optimal allocation of resources is to liberalize the energy sector as much as possible. As we will see in chapter 7, the European Union thus adopted a series of directives to dismember existing monopolies and deregulate energy markets, while the United States recently promoted the rise of independent power producers (PURPA), and France experienced major institutional changes in the gas and electricity sectors in the first decade of the 21st century.

4 The fourth and final problem can be stated as follows: is it better to have specialized or polyvalent energy operators?

   This question may arise as to the exploitation of oil and gas, and deserves to be raised for the distribution of electricity and gas. It obviously arises for energy conservation and new energy sources that present significant synergies.

The answers vary from one country to another depending on their political preferences, their energy situation and their history. At most, we can outline a few points.

First, a political arbitration has to be made between national security requirements and objectives, and cost and efficiency constraints. National security concerns will likely lead some countries to develop the domestic production of previously imported energy, which generally has a positive impact on employment, but they will also lead them to diversify the geographical origin of foreign supply chains, although some may prove more expensive than others. We have seen how these choices interact with the pricing policy.

Conversely, the optimization of costs can lead a country to favor imports, and to focus on the most advantageous supply in financial terms: the advent in the 1960s of what has been called in many OECD countries the "all oil" policy, this oil being imported from the Middle East, is an illustration.

The first choice leads to the development of national companies, the second lets market forces play their part. Obviously, there is a tradeoff between the two constraints: one cannot accept to pay any additional costs for security reasons any more than one can fully sacrifice security imperative on cost. Depending on whether a country is more in favor of one or the other of these options, its energy structures are directed differently. Remember also that these structures reflect the weight of technological constraints, as we previously saw.

Geological considerations also have a decisive influence on the organization of the energy system. The vast majority of countries that have substantial energy resources and a large domestic energy production tend to impose, sooner or later, the emergence of national companies and to limit the intervention of foreign operators.

Finally, we must take into account economic considerations, particularly related to the existence of a national private industry. The failure of private initiative may stem from the difficulty for an operator to raise funds of sufficient magnitude. It can also stem from the degree of risk associated with a geological prospecting operation, or from the political risk of a full forfeiture of the mining income or of a nationalization. Historically, it has often led to an increasing intervention of foreign multinational companies, and then, in a second phase, of the national State itself.

In some other countries, on the contrary, the dynamism of private initiatives has developed a domestic industry without recourse to the intervention of the State: such is the case in the United States and it has long been true for Britain, although the British government has had to create a public company, the BNOC, to exploit the North Sea deposits.

5.6.3 Relationship between the central administration and operators

Relations between the State and energy operators cover essentially three major aspects: regulation, national energy planning and direct supervision of operators.

5.6.3.1 *Regulations*

The questions that arise are why and to what extent the State must regulate the energy sector, and how it must do so.

1 The first aspect of regulation is the existence of a mining legislation. States are generally reluctant to relinquish the control of sub-national soil[1]. They have to regulate the exploration or exploitation of the latter so as to reserve the right to grant or refuse permits. Mining codes thus constitute one of the foundations of national sovereignty in the field of energy.

2 A state may, however, go beyond the mere legal management of granting mining titles, and introduce a regulation on supply. This regulation may concern the domestic supply, for example by setting the market share of the various operators, or the foreign supply, subjecting it to import authorizations.

An additional aspect of this regulation concerns the obligation for operators to hold a minimum safety stock of petroleum products equivalent to three months consumption (see section 2.4.1.5).

3 The regulation of energy prices is inseparable from the overall economic context, including the pricing regime which is in force, but it can also be subject to a specific approach. It is then, in most cases, dictated by two considerations:

- controlling the price changes, so as to prevent the consumer from experiencing erratic changes in a poorly regulated market;
- in those cases where a few operators dominate the market, it should be out of the question to let them set their own prices.

In both cases, the regulation of energy prices reflects a concern for consumer protection. We mentioned the risk of distortion above.

4 Tax regulation is the last major component of energy regulation.

A special tax for the petroleum sector exists in many countries. This regulation has two essential characteristics:

- it has significant financial implications, since the energy taxes account for a high percentage of the State's revenues, whether the country is exporting or importing it;
- energy taxation has a structuring influence:
  - on the energy supply, the purpose of oil companies taxation being to encourage them to take risks;
  - on the demand, since taxes on petroleum products enable the State to recover a portion of oil revenues, to curb the consumption of petroleum products and to prompt the user to use an alternative energy when possible.

In many countries, there also are parafiscal taxes whose raison d'être is to fund specific policies: oil exploration, rational use of energy investment, or building of electricity distribution networks in rural areas.

1 The law of the United States is the exception.

5.6.3.2 *National energy planning*

National energy planning can also be an excellent instrument for the action of a state vis-à-vis its national operators. As this point has already been dealt with in section 5.2, we will not discuss it further here.

5.6.3.3 *Exercising of supervision*

Supervising energy operators is the last aspect of the relations between the State and the energy sector. The question that may arise is whether energy players should be given full management autonomy, or, conversely, whether they should be considered an instrument for the national planning.

The answer to this question obviously depends on the political will of the State, and is different depending on the type of operator in question: the exercise of administration supervision cannot take the same form or the same intensity for a monopoly similar to a public service, a public company operating in a competitive sector, a national operator or a private operator subsidiary of a foreign group.

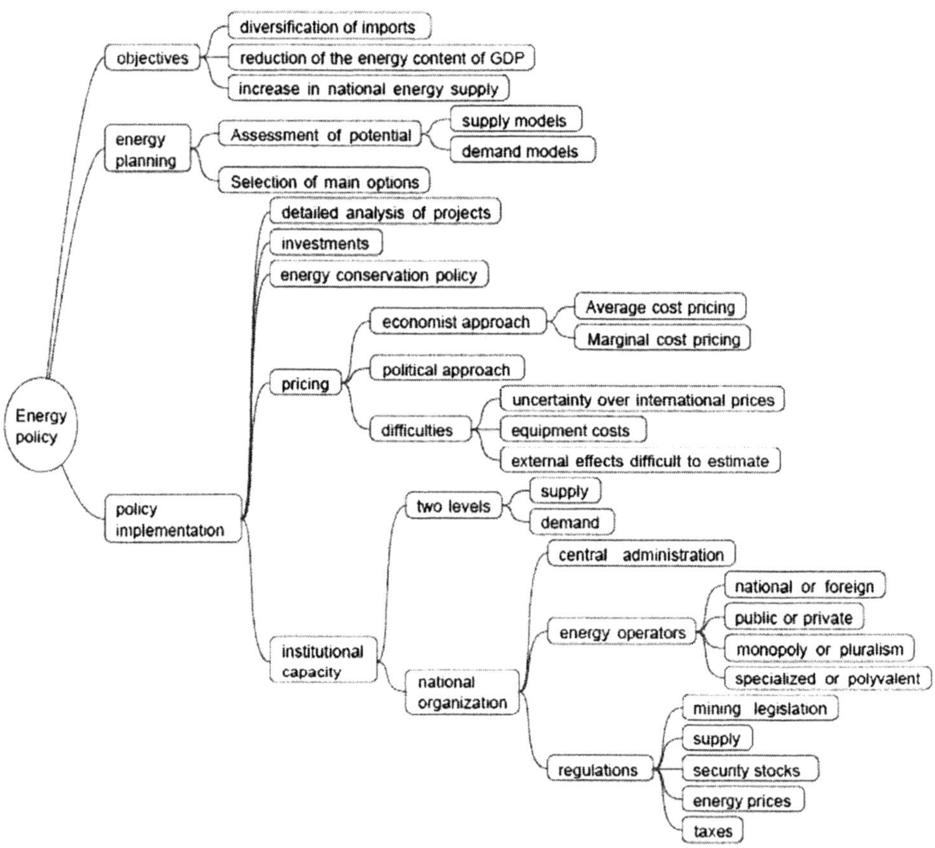

Figure 5.6 Energy policy mind map.

Between close control of the State and almost complete management autonomy, one can find a whole range of possible degrees of intervention.

Public supervision may also be exercised in varied ways: dialogue between the administration and operators, approval by the State of energy companies' plans, agreements based on reciprocal commitments, appointment of officers, presence of state representatives on the boards of the operators, etc.

5.7 CONCLUSION

The mind map in Figure 5.6 gathers the main concepts pertaining to energy policy.

REFERENCES

Eurostat, L'énergie, L'Europe en chiffres – Annuaire Eurostat 2008.

Zagamé P., Modèles de l'énergie et nouvelles théories du progrès technique, Conseil Français de l'Énergie, Paris, décembre 2008.

FURTHER READING

Chevalier J.-M., Barbet Ph., Benzoni L., Économie de l'Énergie, Presses de la Fondation nationale des Sciences Politiques et Dalloz, Paris, 1986.

Conkling R.L., Energy pricing: Economics and Principles, Springer, 2011, ISBN: 978-3-642-15490-4.

Goldthau A., Witte J.M., Global Energy Governance: The New Rules of the Game, Kindle Edition, 2010.

Percebois J., Économie de l'Énergie, Economica, 1989.

Chapter 6

World energy situation

6.1 PRESENTATION

This chapter is particularly important for the study of global energy problems. It is composed of three main parts.

The first outlines energy consumption and production worldwide. Knowledge of the volume of energy consumption in the world and its evolution is indeed necessary to assess the overall trends in the short and medium terms and to understand the pressure that this demand may put on resources.

Examining regional disparities, both in terms of consumption and production, provides the essential elements for a reflection on global geopolitical issues. The second part thus deals with reserves and resources, including oil, which is one of the basic variables of the global geopolitics of energy, as it determines for a large part the policies of exporting and importing countries and their relations with one another in the long-term.

The third part presents some problems specific to developing countries, related to their economy and to both their commercial and non-commercial energy consumptions.

6.2 CONSUMPTION AND PRODUCTION

6.2.1 Primary energy consumption in the world

Approximately 7 billion people lived on Earth at the end of 2011. The total consumption of commercial primary energy was at that time more than 12 billion toe (12 Gtoe), and growing steadily, as shown in the graph in Figure 6.1, despite some minor shifts following the first two oil shocks.

6.2.2 Historical trends

Tables 6.1 and 6.2 show, in million tons of oil equivalent and percentage, changes in consumption of commercial primary energy in the world from 1925 to 2007.

These data prompt the following remarks:

- the global commercial energy consumption has grown at a very steady 2.3% rate per year on average between 1925 and 1950, 4.9% between 1950 and 1973, 2.1% from 1973 to 1988, and 1.8% since;

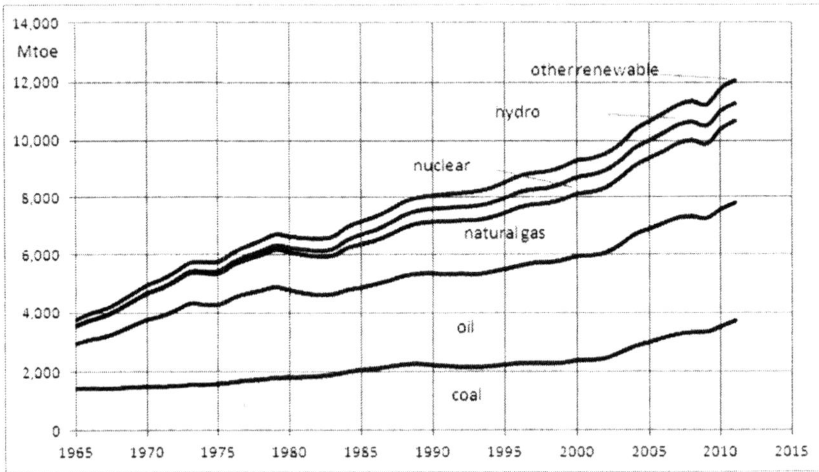

Figure 6.1 Primary energy consumption by source.

Table 6.1 Primary energy consumption in the world.

Mtoe	*1925*	*1950*	*1973*	*1988*	*2007*
Oil	148	542	2 798	3 038	4 001
Gas	36	189	1 066	1 631	2 498
Coal	922	1 195	1 668	2 428	3 206
Hydraulic	7	32	332	537	265
Nuclear			49	439	709
Total	1 113	1 958	5 913	8 073	10 678

Table 6.2 Distribution of primary energy consumption in the world.

%	*1925*	*1950*	*1973*	*1988*	*2007*
Oil	13.3	27.7	47.4	37.6	37.5
Gas	3.2	9.7	18	20.2	23.4
Coal	82.9	61	28.2	30.1	30
Hydraulic	0.6	1.6	5.6	6.7	2.5
Nuclear			0.8	5.4	6.6

- fossil fuels account for nearly 90% of the total, thereby playing a leading role at the global level;
- the share of coal declined significantly for 60 years, from almost 83% in 1925 to about 28% in 1973. It now fluctuates slightly around an average value slightly under 30%;
- the main coal substitution has been provided by oil, whose share has more than tripled from 1925 to 1973, before decreasing and then stabilizing in the 1980s;
- in parallel, natural gas has grown at an equivalent or higher rate, continuing after 1973 and beginning to drop off after 2000;

- in the 1970s, a significant contribution was provided by nuclear power, which developed at a very high rate (16% per year). However, this growth has slowed considerably in recent years;
- finally, the absence of so-called new energies from this table should be noted: they still play a secondary role as a commercial energy.

Figure 6.2 shows that the growth of global electricity production has continued over the last twenty years, regardless of changes in oil prices, and continues today at a rate of around 3%. In addition, since 1950 there has been a very strong growth in nuclear and hydraulic primary electricity, and more recently in wind and solar energy.

6.2.3 Distribution of per capita consumption

As shown in the graph in Figure 6.3, in 2011, energy consumption per capita varied significantly around the world (from 140 koe/inhabitant in the poorest countries

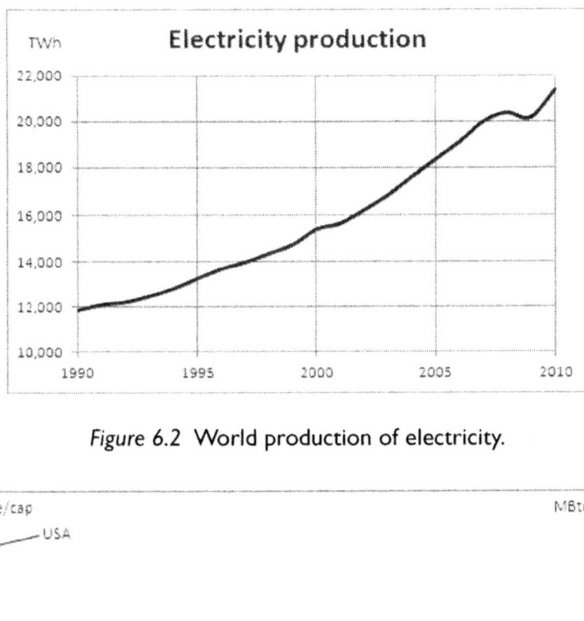

Figure 6.2 World production of electricity.

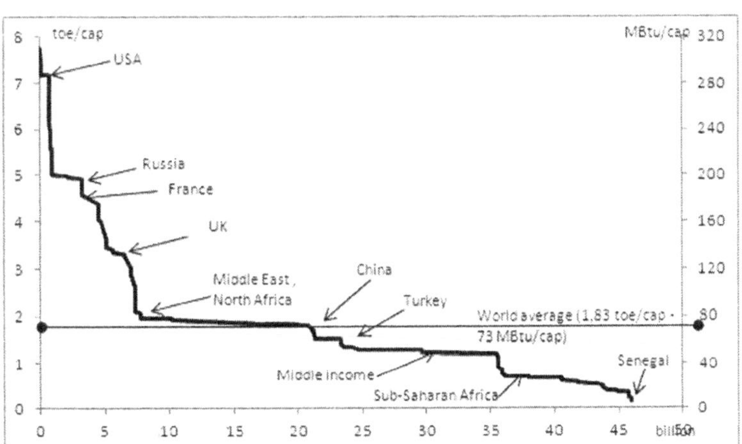

Figure 6.3 Distribution of per capita consumption.

up to 7.2 toe/cap in North America and even 16.9 toe/cap in Iceland), following inequalities in per capita income. The left scale is in tonnes of oil equivalent per capita and the right scale in MBtu per capita.

At a regional level, we see that the distribution of energy consumption is very uneven: in North America, it is about 4 times the world average (1.83 toe/capita) whereas in Sub Saharan Africa it was 2.5 times lower.

In Russia and Europe, the per capita consumption levels ranges between 3 and 4.9 toe/capita.

Some oil exporter developing countries achieve energy consumption per capita similar to those of industrialized countries.

Two major groups of countries are commonly distinguished:

- the group of Industrialized Countries (ICs), which includes about thirty countries with a population of 1.2 billion people, or 18% of the total. Its primary energy consumption was 6.3 Gtoe in 2008, or 57% of the total;
- the group of Developing Countries (DCs), which includes the remaining 150 countries and 5.7 billion people, or 82% of the total. Its primary energy consumption in 2008 was 4.7 Gtep, or 43% of the total.

Remember that these figures only include commercial energy. Non-commercial energy, such as firewood, energy of draft animals, etc. is very difficult to estimate accurately. Values for this kind of energy ranging from a few percent for industrialized countries to about 30% for developing countries are generally accepted. However, these averages can mask very significant differences from one country to another.

Particularly in the Sahel countries, the share of firewood can reach 90% of the total primary energy balance. These figures may seem surprising, but they reflect very low conversion efficiencies (a few percent) obtained by cooking with "three-stone" wood fire, which is a rudimentary fireplace consisting of three stones arranged in a triangle between which wood burns, with a pot being placed above them.

6.2.4 Regional distribution

Table 6.3 shows the distribution of primary energy consumption by region.

It prompts several comments:

- the share of China and the rest of Asia has grown strongly, doubling in 20 years' time;
- the share of Europe in the global energy consumption decreased from nearly 40% in 1988 to 30% in 2008;
- the weight of North America in the global energy consumption has been and remains very important;
- some continents consume a very small part of the total, e.g. Africa, which represents about 3% of global energy consumption.

Note that energy consumption in developing countries has increased substantially, even after the oil shocks.

There is also a huge disparity, highlighted earlier in this chapter, between situations in industrialized and developing countries.

Table 6.3 Primary energy consumption by region.

%	*1925*	*1950*	*1973*	*1988*	*2008*
North America	51	49.4	37.8	27.2	24.6
Western Europe	35.2	22.6	19.9	16.2	15.1
Eastern Europe	5.5	18	23	25	12.6
Japan	2.1	1.8	4.5	5	4.8
ICs Subtotal	*93.8*	*91.8*	*85.2*	*74.7*	*57.1*
Latin America	1.7	2.6	3.9	5.3	7
China	1.6	1.7	5.4	9	19
Asia (other countries)	2	2.3	3.9	8.4	13.6
Africa	0.9	1.6	1.6	2.6	3.3
DCs Subtotal	*6.2*	*8.2*	*14.8*	*25.3*	*42.9*

6.2.5 Production and consumption by region

Table 6.4 summarizes, for the year 2011, primary energy consumption and production of fossil fuels (88% of total) in the main regions of the world (source: BP statistical review 2012). Values are given in Mtoe. To approximately convert them into quads, one can divide them by 25 or multiply them by 0.04.

Note that, in the same way that energy consumption differs greatly depending on the considered region, the production of fossil fuels is highly variable from one region to another, and also depends on fuels:

- developing countries produce just over 60% of the world's oil, with half of it coming from the Middle East;
- however, more than 60% of natural gas production is provided by the industrialized countries, nearly half of which by Europe and Russia;
- Japan and Western Europe are the regions most dependent on external energy supplies, and they were recently joined by China, which is increasingly importing oil;
- Eastern Europe is the main producer of fossil fuels, with Russia producing 511 million tons of oil and 380 Mtoe of natural gas.

Also note that the three biggest consumers of oil are the United States (750 Mt), Western Europe (700 Mt), and Russia (500 Mt). China's (and to a lesser extent India's) oil needs are however still increasing.

Given the impact of these large consumers on the global demand for this strategic product, an analysis of the probable evolutions of their oil production and consumption is essential to study the oil market prospects.

6.2.5.1 *Dynamics of large oil importers*

The graph in Figure 6.4 illustrates the dynamics of major oil importers (the United States, Europe, the rest of the OECD, India and China). The left scale is in million tonnes and the right scale in barrels per day.

Table 6.4 Energy consumption and production by region.

Millions toe 2011 data	*Primary energy*	*Oil production*	*Natural gas production*	*Coal production*
Middle East	748	1301	363	1
Africa	385	417	99	147
Latin America	643	380	139	65
Asia	1 712	185	414	730
China	2 613	204	118	1 956
Developing countries	6 101	2 487	1 133	2 899
Japan	478			7
Europe–Eurasia	2 238	327	609	157
Russia	686	511	382	300
North America	2 773	670	782	600
Industrialized countries	6 175	1 508	1 773	1 064
Total	12 274	3 996	2 906	3 955

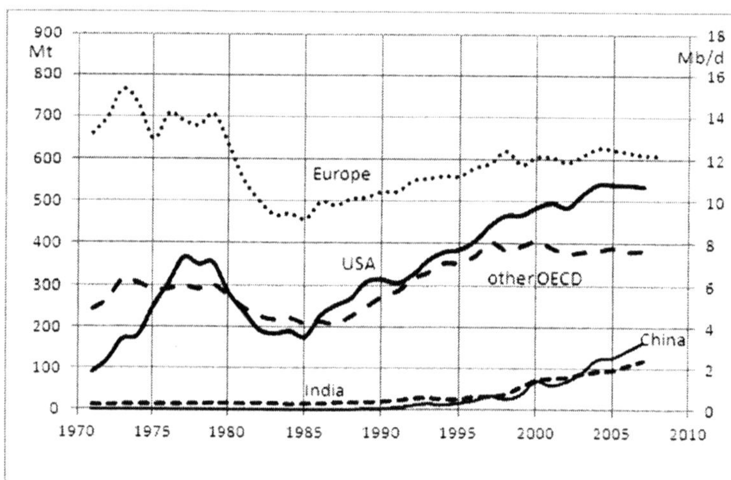

Figure 6.4 Large oil importers.

It shows that these two latter countries have continued to increase their imports despite higher oil prices since 2005, and that the United States very recently reduced them, while the rest of the OECD, Europe included, has stabilized them over the past ten years.

The graphs in Figures 6.5 and 6.6 illustrate the dynamics of production and consumption of oil in the United States and Europe. The increase in the U.S. oil deficit from 1985 to 2005, which resulted in both an increase in demand and a decline in domestic production, has now stopped due to the development of oil shale production.

Europe is more dependent on imported oil than the United States, but it seems to have succeeded in limiting consumption growth, which is still lower than before the second oil shock.

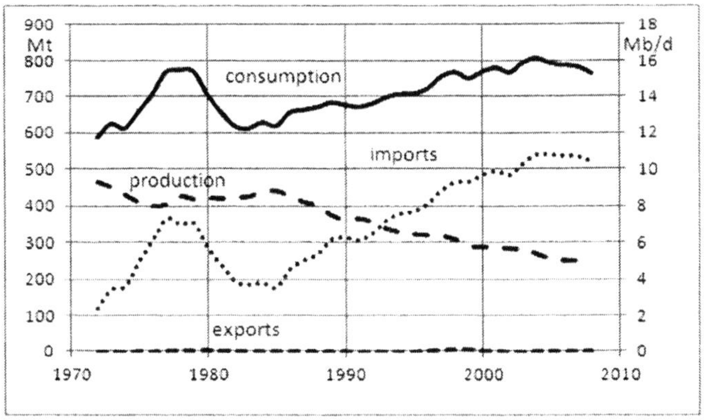

Figure 6.5 United States oil balance.

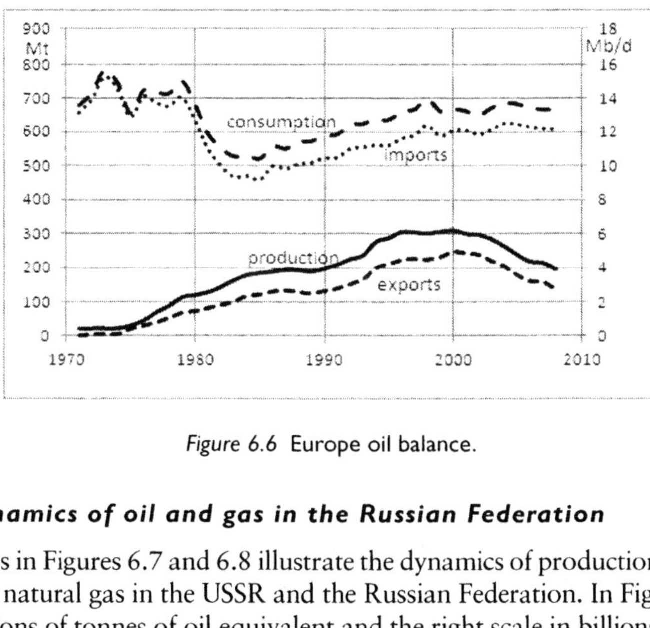

Figure 6.6 Europe oil balance.

6.2.5.2 *Dynamics of oil and gas in the Russian Federation*

The two graphs in Figures 6.7 and 6.8 illustrate the dynamics of production and consumption of oil and natural gas in the USSR and the Russian Federation. In Figure 6.8, the left scale is in millions of tonnes of oil equivalent and the right scale in billions of cubic feet.

The drop in the curve in 1990 stems from political changes at this time. These graphs show that exports are tending to stabilization in both cases, after a period of growth for oil.

6.2.6 Production and consumption by country groups

6.2.6.1 *Typology of countries*

The World Bank identifies four main groups of countries, in order of increasing GNI per capita (2008 values):

- low-income countries (less than \$975 per capita);
- lower-middle-income countries (between \$975 and 3,855 per capita);

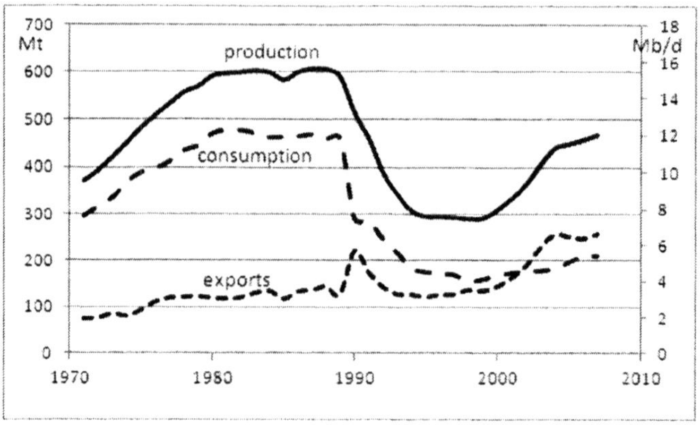

Figure 6.7 Russia oil balance.

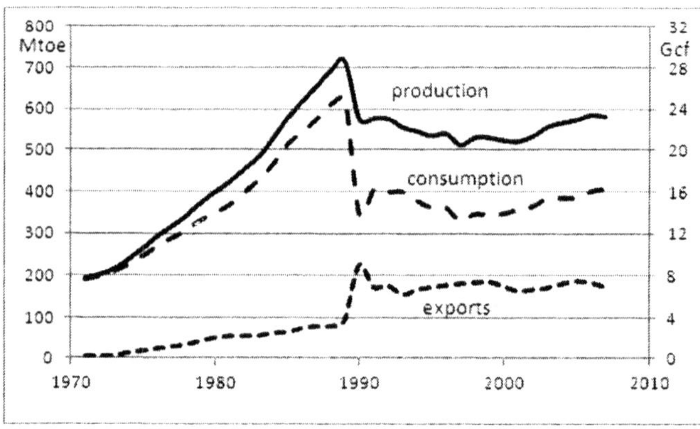

Figure 6.8 Russia gas balance.

- upper-middle-income countries (between $3,855 and 11,905 per capita);
- high-income countries (above $11,905 per capita).

World Bank categories being little used by non-specialists, it is generally considered necessary, to reflect the diversity of situations, to retain at least three subsets among developing countries:

- the least developed countries, or LDCs, are the poorest among the low-income countries. The need to gather LDCs in a differentiated group emerged in the early 1970s. They are heavily dependent on the Official Development Assistance (ODA) they receive, which may represent, for some, up to 90% of their outside revenues. The list, defined by UNCTAD, includes 49 countries, 32 of which are in Africa;
- at the other end of the scale are the newly industrialized countries, or NICs, experiencing very high GDP growth rates, and representing serious competition

for industrialized countries, as they are in the process of catching up to their standards of living (these include Brazil, China, India, Indonesia, Turkey, etc.);
- between these two categories is a set of countries that remains quite heterogeneous but nevertheless share enough common features to be grouped together.

6.2.6.2 *Consumption and income level*

The graph in Figure 6.9 shows the link between per capita commercial energy consumption and per capita income in the world for the year 2010. Vertical lines represent GDP/cap for the different categories of countries and horizontal lines depict their per capita energy consumption. Note that Figure 4.1 shows a similar graph, the axes simply being reversed.

- in low-income countries, average GDP/cap is \$1,140; they account for 800 million people. Their average per capita consumption of commercial energy is 365 kg of oil equivalent (koe) per year;
- in middle-income countries, GDP/cap is \$5,935, while they represent about 5 billion people. Their average consumption of commercial energy is 1,310 koe/cap;
- in high-income countries, GDP/cap is \$33,185, energy consumption is 5 toe/capita, and their population 1,130 million.

Also shown in this graph is the world average energy consumption (1,850 koe/cap).

Note that the disparity in per capita energy consumption follows quite closely that of per capita income, the point cloud pointing along a smooth curve with a relatively low dispersion. There are wide disparities between the consumptions of the poorest countries and those of industrialized countries, some LDCs having very low per capita consumption.

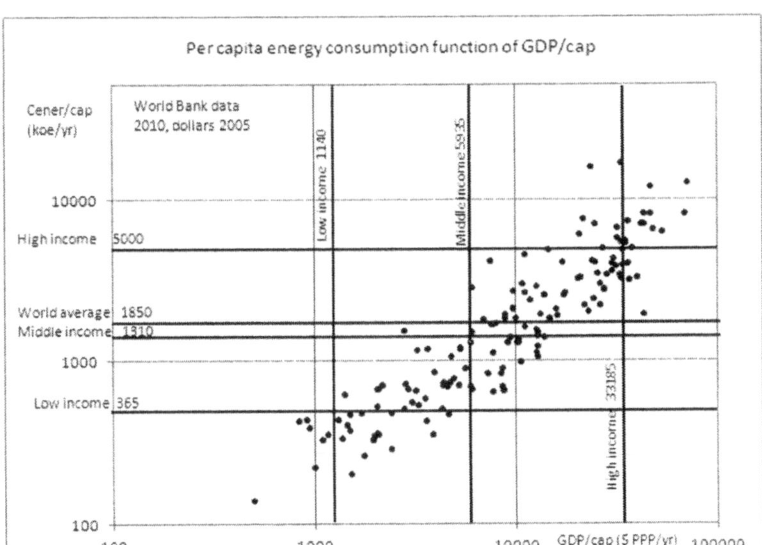

Figure 6.9 Energy consumption and income level.

This graph confirms the existence of a continuity in the development stages, as the scattering is clearly not broken down into distinct subsets.

The graph in Figure 6.10 shows the evolution of per capita consumption in terms of GDP per capita between 1985 and 2010 for the main groups of countries (the left scale is in grams of oil equivalent and the right scale in kBtu). Given that GDP/capita of country groups have generally increased during this period, the horizontal axis also indicates the passage of time.

This graph shows that the strongest increases in per capita consumption took place in middle-income countries and non-OECD high-income countries. In low-income countries and developed countries, per capita consumption remained stable, with some fluctuations in the former countries.

6.2.6.3 *Energy intensity*

Similarly, examining the energy intensities of different countries in the world does not allow us to draw definitive conclusions on a separation between different disjointed subgroups.

Extensive training in energy planning has been implemented over the past thirty years to sensitize policy makers to the different options available in all countries, particularly developing ones, and help them build an energy database without which no action on demand is feasible. These programs have often been accompanied by a change in the administrative and operational structures in charge of the energy sector.

The graph in Figure 3.6 shows the values of energy intensity in 2010 for a large number of countries (in kilograms of oil equivalent per 2005 PPP dollar) in terms of

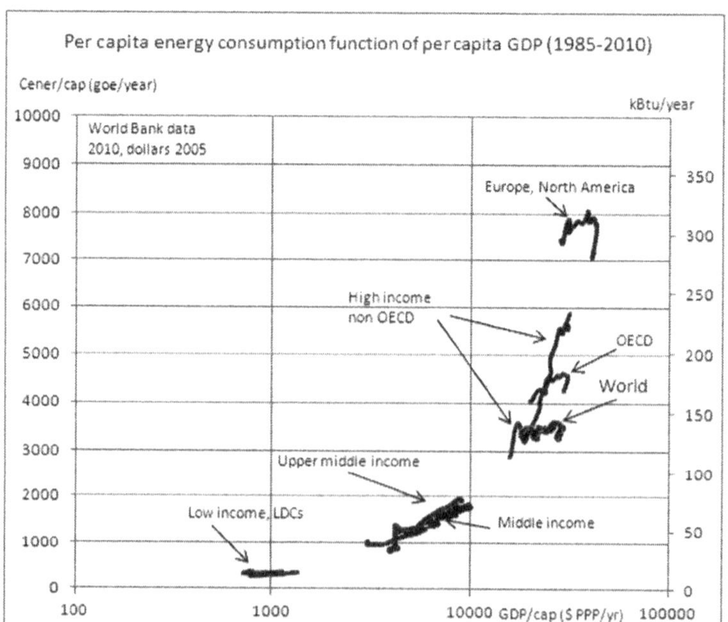

Figure 6.10 Trends in per capita consumption (1985–2010).

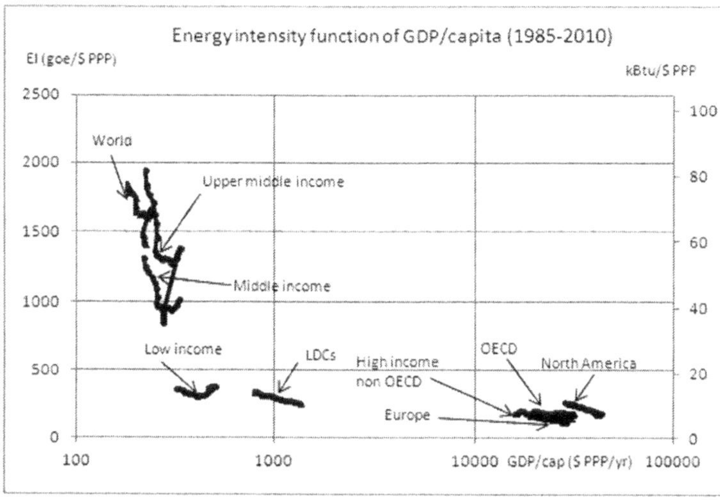

Figure 6.11 Trends in global energy intensity (1985–2010).

GDP per capita. The graph in Figure 6.11 shows the evolution of their values between 1985 and 2010 for main country groups (the left scale is in grams of oil equivalent per dollar of constant PPP 2005 and the right scale in kBtu per dollar of constant PPP 2005). As for Figure 6.10, the horizontal axis also indicates the passage of time.

This graph shows that the strongest declines in energy intensities occurred in middle-income countries, sometimes with variations, but with a constant downward trend direction. In developed countries, energy intensity continued to decline, but at a much slower pace than those of other countries.

There was an increase of energy intensity in some low-income countries, probably due to the construction of infrastructure, but it was, with few exceptions, lower than those of the other groups of countries.

6.3 ENERGY RESERVES AND RESOURCES

6.3.1 Definitions

The concepts of reserves and resources need to be carefully defined. Indeed, these two terms are used in the energy field with meanings that differ quite significantly from common sense, which inevitably leads to confusion and misinterpretation.

"Reserves" is an essentially dynamic concept that varies over time depending on the one hand on economic conditions and on the other hand on exploration and production activity.

The notion of "resource" is more blurred: it includes not only future production but also increased production (ultimate resources).

The only available and reliable information about them concerns the proved reserves, which are the "part of the resources that are both identified and can be produced in the technical and economic conditions of the time", while resources are "those quantities which are believed to exist and could be discovered one day". The latter is a purely geological concept that does not take into account technical or economic constraints.

Note now that the methods for estimating reserves are not standardized, so that the figures announced by countries or companies are not always reliable, far from it, especially since they obviously play an important role in the estimation of their financial assets. This is why the reported figures may vary from one year to another, as they often did in the past.

The *R/P* ratio is the ratio of proved reserves to annual production. It gives an indication of the lifetime of the reserves.

Two points need to be made regarding these definitions:

- proved reserves may vary, both according to price, rates of technically feasible extraction, discoveries of new deposits and rate of production. Thus, for several decades, the oil reserves in the United States have been estimated to be about 10 years of domestic production ($R/P = 10$);
- given their definition, resources' estimations can be quite arbitrary ("quantities which are believed to exist"). Various methods exist to estimate them, mainly based on "reasonable" extrapolations of the discoveries made in the past.

6.3.2 Reserves of oil and natural gas

To illustrate the dynamic nature of the reserve concept, the diagram in Figure 6.12 shows the variation of estimates for oil and natural gas reserves over the last thirty years (the left scale is in billion tonnes of oil equivalent and the right scale in billion barrels of oil equivalent).

6.3.3 McKelvey diagram

Given the importance of these dynamic aspects, it is customary to plot the reserves and resources on a McKelvey diagram (see Figure 6.13), where the quantities are represented by rectangles delimited by horizontal or vertical moving boundaries, which vary depending on the findings, production and economic conditions.

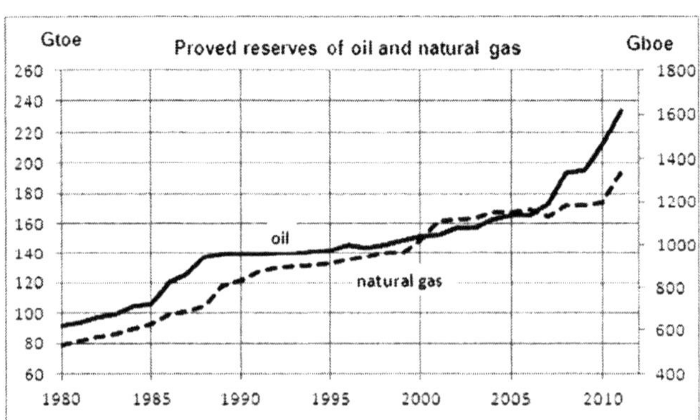

Figure 6.12 Proved oil and natural gas reserves.

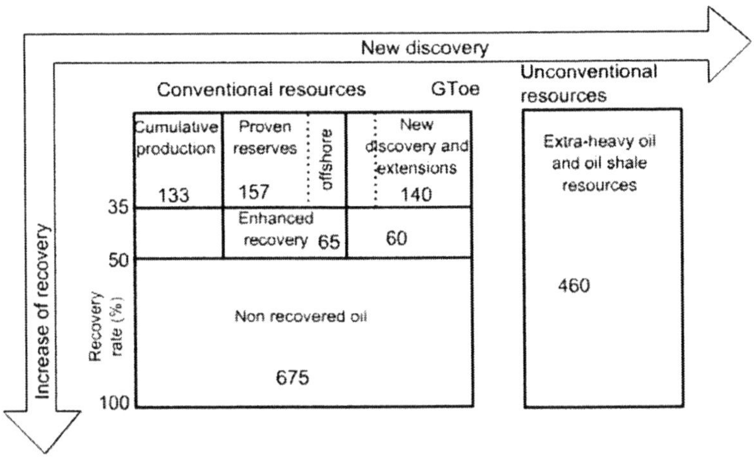

Figure 6.13 McKelvey diagram.

The horizontal axis corresponds to the geological knowledge and the vertical axis (downward) to the profitability of production.

The horizontal boundary separates the technically and economically feasible total production (past and future) from energy that is not recovered. Given the uncertainties concerning recovery in the future, it is often split to reveal the zone corresponding to the recovery (between 25% and 40%).

Vertical boundaries define the area of identified resources, i.e. those that are assured, and resources to identify, the uncertain ones. Among these resources, only the technically and economically recoverable share is likely to fall within the category of reserves.

The diagram in Figure 6.13 was established in 2007 by the French Petroleum Institute.

6.3.4 Proved reserves

In 2011, the world's proved oil reserves were estimated at 234 billion metric tons or 1653 billion barrels, that is to say a little more than 54 years of production (*R/P*).

A characteristic of oil reserves is their uneven geographical distribution (Figure 6.14):

- 14.3% of the reserves are located in North America, with a *R/P* ratio of 42 years;
- 3% in Europe and Eurasia;
- 2.3% in Asia Pacific;
- 5.2% in the Russian Federation;
- 7.5% in Africa;
- 21.5% in Latin America;
- almost 46% in the Middle East with *R/P* = 65 in Saudi Arabia.

Proved reserves of natural gas were estimated in 2010 to be 175,000 billion m^3 (175 Tm^3) or 6,200 billion cubic feet (6200 Tcf), i.e. about 160 billion toe, that is to say a little less than the oil reserves (Figure 6.15).

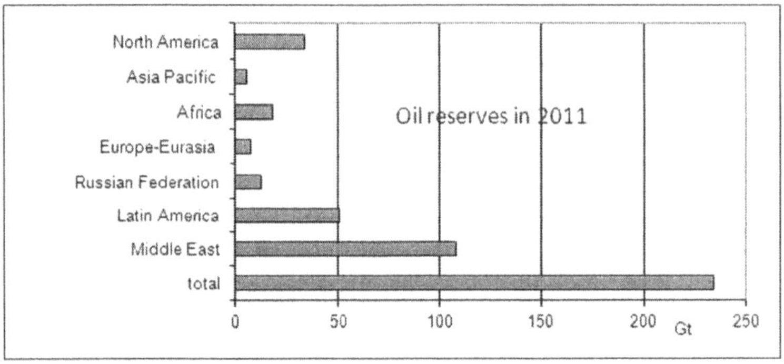

Figure 6.14 World oil reserves.

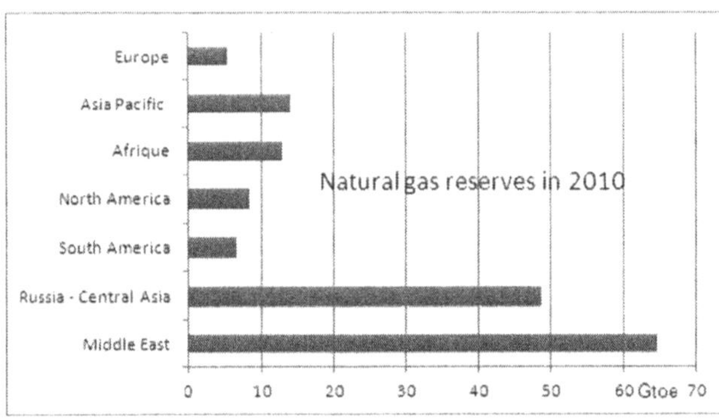

Figure 6.15 World reserves of natural gas.

The geographical distribution of these reserves is as follows:

- 30% in Russia and Central Asia;
- just under 40% in the Middle East;
- 5% in North America, 3% in Europe, Africa 8% (2.4% in Algeria), 4% in Latin America and 8.6% in Asia Pacific.

In 2010, the global *R/P* ratio for natural gas was approximately 55 years [Mathieu, 2010].

In estimating coal reserves, there are generally three qualities to consider: coal, sub-bituminous coal and lignite, each corresponding to a different heating value.

Proved coal reserves were estimated to be 861 Gt in 2011, with a *R/P* ratio equal to 112 years, of which 405 Gt of coal (Source: BP statistical review 2012). Considering an average of 0.51 toe equivalence per tonne of coal, proved reserves are about 440 Gtoe.

Their geographical distribution is as follows [Bessereau & Saniere, 2010]:

- 29% in the United States;
- 19% in the Russian Federation;
- 14% in China;
- 9% in Australia;
- 5% in Europe.

In 2008, uranium reserves at a cost of less than \$130/t were estimated by the OECD to be about 3.3 Mt. They are fairly well distributed throughout the world, the main deposits being located in Australia (23%), Kazakhstan (15%), North America (14%) and Africa (21%).

World production is between 40,000 and 60,000 t/year (40,000 t in 2007), for a consumption bordering on 67,000 t/year. Uranium reserves, with current nuclear reactors, therefore correspond to about 80 years of power generation, an equivalent of 48 Gtoe [Gabriel & Saniere, 2010].

Figure 6.16 summarizes the global reserves of energy for the technology available in 2011.

6.3.5 Resources

Today, oil recovery rates rarely exceed 40%, and it is estimated that technical progress allowing a 1-point improvement of this factor would result in 5.5 Gt additional reserves, i.e. nearly two years of the current world consumption.

Estimates of oil resources vary greatly according to the authors. They are probably less than 400 Gt or 20 Tb [Mathieu, 2010].

The natural gas resources were estimated to be about 245 Gtoe in 2010, with a range from 230 to 280 Gtoe (9500–9900 Tcf) [Mathieu, 2010].

The usual recovery rate being 70–80%, the impact on reserves of technical progress in the extraction is much lower than in the case of oil.

Resources in unconventional hydrocarbons are very difficult to estimate. Shale resources are very significant (estimates vary from 500 to over 2000 Gtoe, more than 400 having been measured). Given the uncertainties about the technologies and their

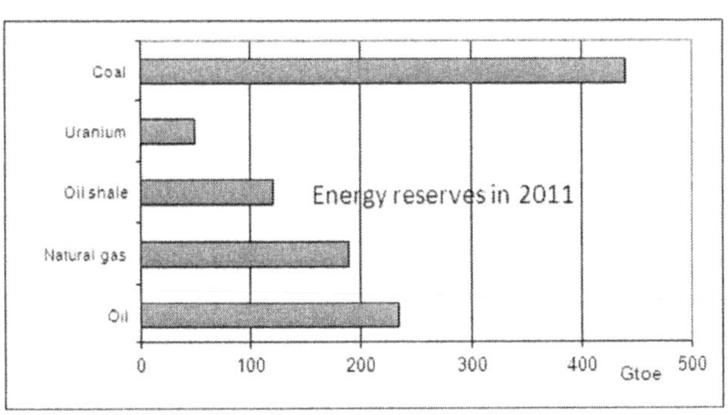

Figure 6.16 World energy reserves.

impact on the environment, it is not possible to know today what percentage of these resources will actually be recovered.

We present here the estimates published by the IFP, based on data provided by the IEA [Vially, 2012]. Ultimate resources of liquid fuels (Figure 6.17) would be close to 9,000 Gb or 1230 Gtoe, and those of gaseous fuels 900 Tcm or 840 Gtoe. The estimates of liquid fuels are plotted in figure 6.17 where the ordinate represents the production cost range in $ 2008 per barrel, and the abscissa the cumulated amount of resources (in Gb). The estimates of gaseous fuels are plotted in figure 6.18 where the ordinate represents the production cost range in $2008 per MBtu, and the abscissa the cumulated amount of resources (in Tcm).

Coal resources are considerable. In 1980, they were estimated at 7700 Gt, of which almost 5,500 Gt of coal. It should be noted that these estimates are very imprecise, some countries not having reassessed their resources since the beginning of last century.

Estimates of uranium resources vary in between 11 and 32 Mt.

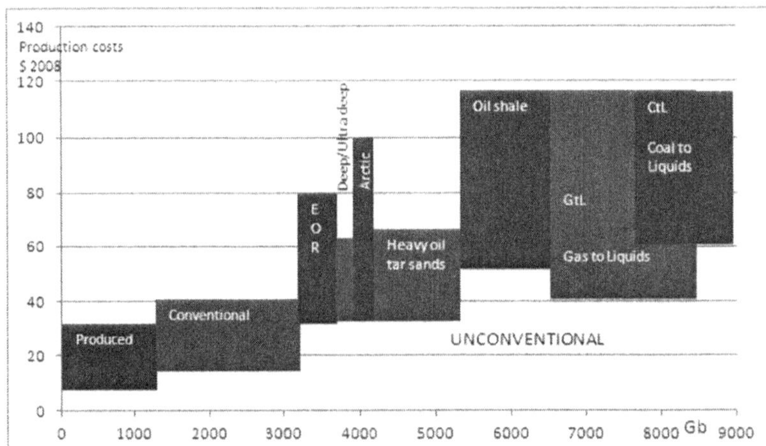

Figure 6.17 Ultimate unconventional liquid hydrocarbon resources.

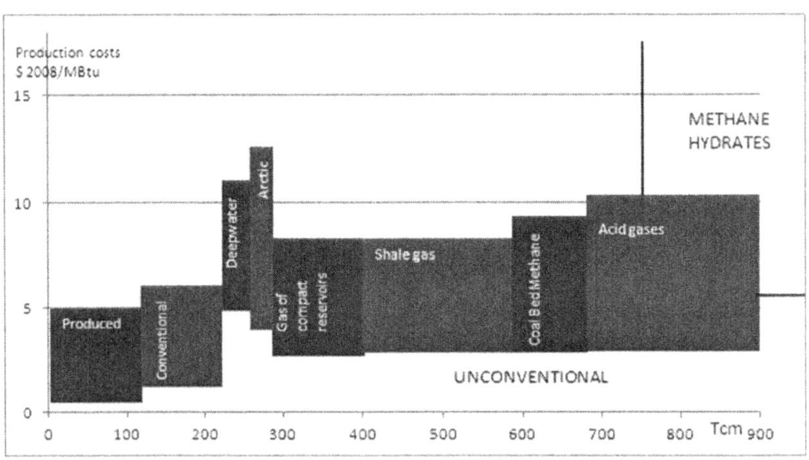

Figure 6.18 Ultimate unconventional gaseous hydrocarbon resources.

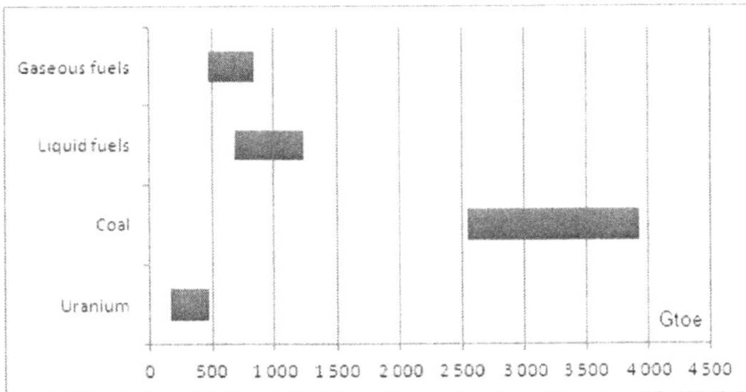

Figure 6.19 World energy resources.

The graph in Figure 6.19 summarizes the ranges of estimates of energy resources. Those of uranium do not account for fast neutron breeders. This graph clearly shows the importance of the availability of coal.

6.4 ENERGY PROBLEMS IN DEVELOPING COUNTRIES

6.4.1 Definition

The term "Developing Countries" (DCs) appeared in the jargon of the United Nations to replace "under-developed countries", its connotation being considered pejorative. It is difficult to define what the phrase "developing countries" exactly refers to, and the easiest way is probably to consider that it refers to the countries that are not industrialized. In section 3.3.5, we have seen that in general their human development index is less than 0.8.

When analyzing the composition of this group of countries, one finds that it is far from uniform, so that it would probably be better to drop the industrialized/developing countries dualism altogether, as it is impossible to draw a clear distinction between these two categories of countries. At least, if one continues to refer to this opposition for the sake of convenience, one should be aware that it is a very rough model for a very complex reality, as we shall see in this section.

If, as we saw in section 6.2.6, the value of per capita income is not sufficient to distinguish DCs from other countries, they do however have share a common characteristic, which is the very high rate of their population growth: about 2% per year, against 0.6% in industrialized countries, although, as we said, very contrasting situations exist in different countries.

This feature means that maintaining living standards already requires the growth of economic activity by 2% on average due to the mechanical effect of the population increase, and that any increase in per capita income growth implies a still greater growth.

Let us, therefore, recall that the concept of "developing countries" covers a very diverse and evolving reality, which cannot be strictly defined, and includes internal discrepancies that may be higher than those that distinguish them from industrialized countries.

6.4.2 Dual societies

One of the most marked characteristics of the group of developing countries is the coexistence, in almost all of them, of a "modern" sector quite similar to that of industrialized countries, and a "traditional" sector that has hardly evolved over the centuries. Very roughly, the modern sector is present mainly in big cities and the traditional sector in rural areas, with small towns constituting the interface between these two types of societies.

The separation between the modern and traditional sectors gives rise to dual societies divided between elites and disinherited masses. Schematically, the elites (less than 10% of the population) maintain close relationships with the rich industrialized countries, and largely mimic their lifestyle, which leads to patterns of consumption well above the average of their country. The great population masses are distributed between rural and city slums, living in the most deplorable conditions, and are kept well away from the few islands of prosperity.

On the energy level, situations in the modern sector and the traditional sector are mixed: on the one hand the economy is based primarily on commercial energy sources, including oil, on the other hand it is still an agricultural and pastoral economy, traditionally based on "free" and locally available energy (firewood, vegetable waste, wind, draft animals, etc.).

We have already indicated that data on non-commercial energy consumption is very unreliable and generally non-existent. We must therefore use the numeric values proposed for them in the literature with caution.

With these caveats in mind, and to illustrate the importance of the role of non-commercial energy in developing countries, let us consider the following figures as representing the percentage of the 2009 population relying on biomass as the main fuel for cooking [IEA, UNDP, UNIDO, 2010]:

- 67% in Africa;
- 80% in sub-Saharan Africa;
- 55% in Asia;
- 32% in China;
- 18% in Latin America.

6.4.3 Oil importers DCs

A distinction must obviously be made among developing countries between exporters and importers of energy (and especially oil), the energy situations of these two groups being very different. Oil-importing developing countries were more strongly affected than industrialized countries by the oil shocks since they could not compensate for the deficit in the trade balance by increasing their sales of manufactured goods. As they do not have the complex energy networks that exist in industrialized countries, they did not have the opportunity to use inter-energy substitution to depart even provisionally from the oil constraint. Oil is the energy source most appropriate to the situation in many developing countries and, very often, the oil shocks have plunged the poorest of them into deep recessions.

6.4.4 Energy problems of the poorest

In developing countries, the poorest mainly use non-commercial energy, as their purchasing power does not allow them to have access to commercial ones [UN, 2012].

Two broad categories should be considered:

- people in rural areas;
- urban dwellers.

Residents of rural areas often live outside the monetized sphere and commercial networks. The importance of rural areas is due to the fact that in 2011 53.5% of the developing countries population was living in such areas, rising to 63.3% in sub-Saharan Africa and 71.5% in LDCs (Source: United Nations). In these countries, despite the strong increase in urbanization in recent decades, the majority of the population still lives in rural areas, their energy supply being based mainly on non-commercial energy, which is used for seeing to domestic needs (cooking, heating and lighting).

The nature of the energy problems encountered strongly depends on the economic environment and energy situation, but also on socio-cultural factors such as traditions, the level of education, etc. For example, the concept of management of forest resources is often alien to nomadic peoples. Most often, energy incursions on the surrounding environment are not given a second thought, and the connection is not always made between scarcity and overexploitation.

For several decades, rural electrification programs have been undertaken in many countries, but they generally involve a relatively small fraction of rural villages. Even in areas that are electrified, the percentage of the population connected to the network is low, considering the standard of living, and the poorest are excluded from modernization.

Cooking remains largely supported by firewood or charcoal, both in culinary tradition and because this fuel is "free", even if its collection entails a considerable amount of work (it is not uncommon for a family member to have to walk several kilometers a day to collect wood and bring it home).

Moreover, poor fuel quality, high humidity and the frequent absence of chimneys mean that combustion takes place in poor conditions and that people inhale toxic fumes. WHO estimated that more than 1.45 million people die each year from this domestic pollution, including many children [IEA, UNDP, UNIDO, 2010].

Lighting is provided by oil lamps, extracted from plants, which are gradually replaced by subsidized kerosene when it is distributed in the village. This form of lighting is also harmful to health because of the fumes it emits, and as it moreover presents a risk of fire and burns.

Agricultural mechanization only exists in a tiny proportion of rural areas in developing countries, local transportation and the mechanical energy necessary to agriculture being provided by labor of human or animal origin.

The inhabitants of cities, including slums of urban peripheries (slums of all kinds) constitute the second category. They generally have a rather low standard of living, but because of their proximity to energy networks and water supply, they have access to commercial energy more than non-commercial ones. Note also that in the cities, the notion of non-commercial energy is misleading in the sense that the firewood and charcoal, which are quite heavily used in some urban centers, are collected in the neighboring forests and distributed through an entirely commercial network. Strictly speaking, this energy thus becomes a commercial energy. In fact, as it is rarely accounted for, one has to continue to consider them, by misuse of language, as non-commercial energy.

6.4.5 The firewood crisis

Often overlooked in developed countries, the firewood crisis is a reality for a significant share of the world's population. Intergovernmental agencies reckon that more than 2.7 billion people use firewood every day as their main source of energy [IEA, UNDP, UNIDO, 2010].

In 1980 it was already estimated that over 100 million people lived in areas with a shortage of firewood in developing countries, and that over one billion were drawing on reserves faster than they were being replenished. In 2012, the site of the World Bank on poverty and deforestation mentioned the figure of 800 million people who depend heavily on these resources, and indicated that 5% of tropical forests are destroyed every ten years.

While the commercial energy crises consecutive to oil shocks experienced cyclical changes, since physical limitations do not constitute a threat in the longer term, the firewood crisis knows no respite because the situation continues to deteriorate in many rural areas, particularly around urban centers.

The annual rate of deforestation in the world is now estimated to be nearly 13 million hectares, against 16 in the 1990s, which shows that governments are beginning to react. The basic reason is that in rural areas, firewood is the most popular fuel for domestic and handicraft needs, because of its safety and the near absence of alternative energy [MDG, 2010].

In cities, the use of charcoal preferentially to firewood reflects the cleanliness of the fuel, its superior heating value (almost double that of firewood), and its ease of transport over long distances and of storage, all despite a relatively low efficiency of carbonization (about 18% in the traditional clay wheels).

Although a predominantly urban fuel, charcoal is used in rural areas in some places of production. Similarly, when wood is relatively abundant, consumption shifts to charcoal for cost issues.

6.4.5.1 International programs

For years, especially under the leadership of FAO and the United Nations, international programs have been actively promoting more efficient methods of production and use of firewood. They relate in particular to:

- increasing the productivity of existing resources through better forest management;
- creating new forests through reforestation programs using high-efficiency species;
- organizing the distribution of firewood;
- improving conversion techniques, either to produce charcoal, or through the design and dissemination of improved stoves (the efficiency of the "3-stone" furnace is in the best cases 8–10% and often much less, and that of the "Madagascan furnace[1]" for charcoal is about 20%).

1 We call Madagascan furnace a traditional charcoal stove made of sheet metal, whose bottom is provided with holes allowing combustion air and ash fall to pass through.

6.4.5.2 *Inconclusive results*

However, it must be recognized that efforts seem trivial given the scale of the problem, as evidenced by the following findings:

- in 1980, the plantations in the Sahel were 32 times smaller than the surfaces deemed necessary;
- the organization of forest management faces considerable socio-cultural difficulties, because the collection of firewood is considered by the inhabitants of rural areas as a levy on natural environment, and not as a form of agriculture. In nomadic areas or on grazing routes the management of timber resources seems impossible;
- the results provided by the "improved" stoves do not seem to live up to the expectations of their promoters, despite some successes. This is mainly due to the difficulties of diffusing technological change in traditional rural societies.

A major element aggravates this context: most affected populations live in a subsistence economy outside the monetized sphere. Their purchasing power is extremely low and optimizing the performance of their stoves under these conditions cannot be their primary concern.

6.4.5.3 *Substitution by LPG*

With all actions relating to the management of forest resources and techniques to use wood unfortunately having had a limited impact, a viable solution in the long term could be the substitution of an alternative energy (particularly LPG), especially in urban centers.

Indeed, on the basis of an average annual wood need of 1 m^3 (470 kg) per capita for cooking (data relative to West Africa), one can estimate, taking into account conversion efficiencies and heating values of various fuels considered that 23 kg of LPG per capita per year would be needed, against 137 kg of charcoal (Table 6.5).

A study conducted in the 1990s in countries of the Mediterranean Basin showed that, with this substitution, one could technically provide a solution to a shortage from which 40–45% of the populations of the concerned countries are going to suffer, by increasing only by 0.5% the total commercial energy consumption, and costing 40 to 80 times less than what an effective reforestation program would. A study by the IEA, UNDP and UNIDO in 2011 estimated that providing 22 kg/year of LPG to 445 million people would simply increase oil demand by 0.9% [IEA, UNDP, UNIDO, 2010].

Table 6.5 Comparison of fuel for cooking.

Fuel	*Conversion efficiency*	*LHV*	*Annual requirements per capita*
Wood	0.1	18 MJ/kg	470 kg (1 m^3)
Charcoal	0.2	33 MJ/kg	137 kg
LPG	0.8	46 MJ/kg	23 kg

Such an experiment conducted in Senegal however showed that socio-cultural resistances must be overcome in order for the substitution to succeed. These resistances are related to the very low financing capacity of affected populations, often unable to raise funds to buy the needed gas cylinder and burner, to cooking habits, fire gas being considered too bright, and to gas distribution, which does not have a pre-existing network, contrary to charcoal.

6.4.6 Commercial energy needs

Many studies have been conducted over the last forty years to find if some development models are more effective than others in order to ensure sustained economic growth in developing countries. One must acknowledge that there is no miracle model, and that the solutions depend mainly on the social and economic context of each country. More fundamentally, no consensus has been reached on what the very concept of "development" means, to the point where economic development does not have a sufficiently coherent theoretical framework to guide the development of proven strategies in this area. One can assert, though, that without significant economic growth development cannot occur, and that this growth requires industrialization.

As we saw earlier, there is a strong historical link between per capita income and per capita energy consumption. Although thirty years of experience in the field of the rational use of energy suggest that it could be possible in the future to reduce individual energy consumptions and thus continue decoupling energy consumption growth and economic growth, it is almost certain that, for a given state of technology, the latter cannot be achieved without additional inputs of energy.

Moreover, industrialization also requires large inputs of commercial energy, although it is hoped that some development strategies could be carried out without a highly energy-intensive industry. Indeed, the construction of the basic infrastructure of a country, such as the development of transport networks, of cities, etc. requires a lot of energy.

Finally, it is increasingly necessary to substitute commercial energy for non-commercial energy to meet basic social needs that are still unmet today (regarding food, education, health, etc.), and to combat the firewood crisis mentioned above.

All of these elements combined ensure that commercial energy needs in developing countries should strongly increase in the future. Recent changes in the global energy mix also clearly confirm this trend, as the energy consumption in developing countries increases by 2.7% per year [BP, 2012].

More specifically, the growing needs cover four key areas:

- transport;
- urbanization;
- industrialization;
- electrification.

6.4.6.1 *Transport*

The need for mobility is as deeply felt in developing countries as it is in industrialized countries, and its fulfillment usually requires an increase in the standard of living. As a result, in most countries, the number of four-wheel vehicles per capita is continuously increasing (Figure 6.20) and could double by 2030 [Dargay et al., 2007]. Vehicle fleets in developing countries are very different from those in industrialized countries, and they include a large number of motorcycles.

Due to the significant decline vehicle prices have undergone over the last fifty years, the number of vehicles per capita is now higher in developing countries than it was in industrialized countries when their income level was similar. One can therefore expect a strong growth in consumption in this sector in the coming years (the World Bank deems it likely that a 5-fold increase in the number of vehicles – including two-wheelers – will occur in developing countries over the next 20 years).

One of the main constraints is that transport has so far represented an almost captive market for oil. Any growth of consumption in this sector is linked to this energy source, with the constraints we discussed above. Indeed, technological advances in industrialized countries in terms of fuel consumption gradually become widespread, but in general, specific consumptions in developing countries are likely to remain higher.

That is why, like many industrialized countries, some emerging countries like China implement very large electric vehicles (including motorcycles) development programs to avoid the use of combustion engines for urban transport. The objective of these policies is threefold: to reduce fuel consumption, but also to reduce pollution in cities and to master new sectors of a promising technology. Brazil meanwhile has favored ethanol since the 1970s and more recently implemented "flex-fuel" fleets (see Section 2.4.4.2).

6.4.6.2 *Urbanization*

The very low income level in rural areas and the attractiveness of large cities generated a large rural exodus, which, combined with the very high rate of population growth in cities, led to rapid urbanization in developing countries. Here are a few figures in order to measure the extent of the phenomenon: between 1965 and 1988 in the

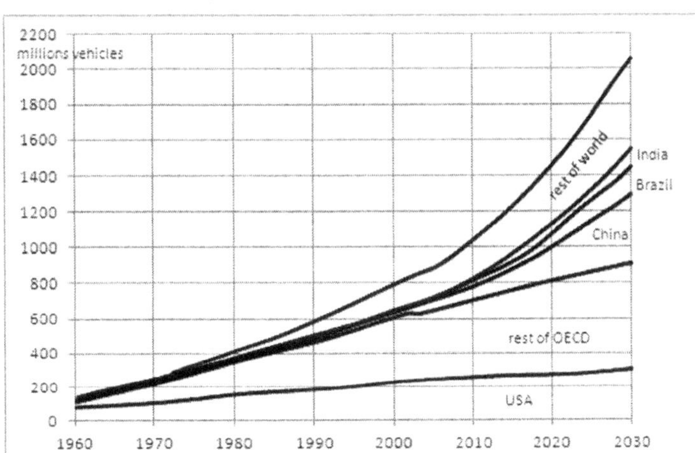

Figure 6.20 World fleet of four-wheeled vehicles.

low-income countries, the urban population increased from 277 million to 1 billion (doubling as a percentage). In middle-income countries, it has grown from 250 to 620 million, while it little increased in the OECD countries (from 443 to 580 million). Since 2008, 50% of the world population lives in cities and this figure will continue to increase in the future. In 2030 cities in developing countries are expected to represent more than 80% of the urbanized world population.

Figure 6.21 shows the proportion of the urbanized world population in the past and estimated by 2050.

With 35 million people, Tokyo is the most populous urban agglomeration in the world, followed by Mexico (18.7), New York (18.3), Sao Paulo (17.9) and Mumbai (17.4).

Despite significant progress in reducing the share of the population living in slums in developing countries, which decreased from 46.1% in 1990 to 32.7% in 2010, their number has increased from 657 to 828 million.

Faced with the growth of cities, politicians are particularly unable to mobilize resources that would meet the needs of urban populations.

Urbanization is accompanied by an energy transition where commercial energy is replacing non-available non-commercial energy. This causes an additional swelling of the growth of commercial energy consumption, beyond energy required for the construction of houses, roads, transportation of individuals, etc. However, the efficiency gains that can be expected due to the use of commercial energy are considerable, so that, in terms of primary energy, the share of the domestic sector is expected to decrease slightly.

In the service sector and in the urban informal sector however, commercial energy consumption grew much faster than in the domestic sector. There was a huge increase in electricity consumption, especially in hot countries where air conditioning can account for a significant share of the total, with the corollary of an oversized installed capacity.

6.4.6.3 *Industrialization*

In many middle-income countries, industry is an important part of energy consumption. Compared with international standards, energy efficiencies are relatively low,

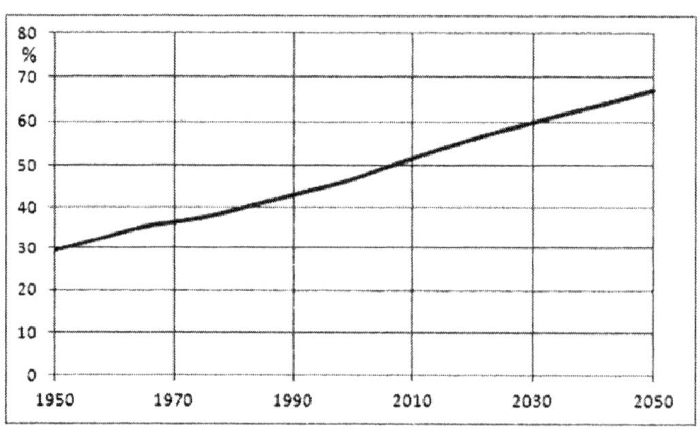

Figure 6.21 Share of world population living in towns.

which suggests to consider the possibility of significant reductions in energy intensity in the future, if new technologies are made available at affordable prices for developing countries.

We have already mentioned above that industrialization appears in almost all countries as a prerequisite on the path of development, and that the construction of industrial infrastructure requires a lot of energy. It is therefore likely that, despite significant technological advances, the share of the industrial sector in developing countries will grow, leading to a significant increase in commercial energy needs.

6.4.6.4 Electrification

Generally speaking, in every country, both developing or developed ones, electricity consumption is growing much faster than income. While the primary energy content of GDP fell in most OECD countries, its final electricity content increased until 1985 and has only begun to decline recently (Figure 3.11). In developing countries, the growth rate of electricity consumption is very high (up to 8–10% per year), many households (almost 75%) still not being connected to a power grid. It is estimated that in 2010 1.4 billion people had no access to electricity, 85% of them living in rural areas [IEA, UNDP, UNIDO, 2010]. The importance of electricity is such for welfare that it is estimated that the development of electricity is a fairly accurate indicator of economic and social progress in developing countries.

Levels of per capita consumption are highly variable, ranging from a few tens of kWh to over 15,000 kWh in some industrialized countries (Sweden), the global average being close to 2,000 kWh.

Given its social impact, rural electrification is often a target of choice for development programs. Electricity indeed brings lighting, television (which provides a first access to information and culture), the possibility of storing medical drugs, it can power small household appliances, etc., all things that contribute to raising the level of immediate comfort.

The main obstacle is the cost of electrification programs, the electricity sector being one of the most capital-intensive sectors. It is common for electrification to represent more than 50% of the total investment budget in the energy sector. As the financial capacity of utilities in developing countries is often limited due to inadequate tariff structures (e.g. because of consumer subsidies), the financing of electrification programs can rarely be provided locally, which requires to use external capital, and therefore increases the external debt. Funding becomes one of the major limitations in the development of the energy sector.

The study conducted by the IEA, UNDP and UNIDO in 2011 which we cited about the firewood crisis also looked at electrification, especially in rural areas, retaining the assumption of minimum annual electricity consumption per capita around 250 kWh/year in rural areas and 500 kWh/year in cities, at the initial time of connection to electricity, and then growing. Providing these modest amounts of electricity to the entire planet by 2030 would lead to an increase in the annual electricity consumption of 950 TWh at that time, i.e. a little less than 3% of total consumption.

The additional investment amounts needed have been estimated to be $33 billion per year by 2030 ($223 billion for the 2010–2015 period, and $447 billion for the period 2016–2030), the main corresponding to decentralized electricity (43% for mini-grid and 20% off-grid). Note that this sum represents only 5% of total investments in the electricity sector.

6.4.7 Energy characteristics of developing countries

In conclusion, energy in developing countries is characterized by:

- an oil crisis, the cost of oil in foreign currency having become excessive for some importing countries, who are forced to seek alternative energy solutions if they do not want to harm their economic growth;
- a crisis of traditional fuels, firewood in particular, directly affecting more than two billion people;
- a strong rural component, which incurs a heavy reliance on non-commercial energy;
- an industry in building, which results in higher energy intensities;
- the lack of infrastructure: due to the lack of networks, oil is the energy source of choice because it can be easily distributed;
- transport that is heavily dependent on oil, which continues to limit substitution possibilities.

In oil-importing developing countries, oil shocks have often had a stronger impact than in industrialized countries, where, consumption being much larger, savings are possible, either by structural effect or by reducing specific fuel consumption.

In the poorest developing countries, development itself can be blocked due to lack of energy, since it is needed to provide the basic necessities for survival or to provide fertilizers, mechanization and water for agriculture, the fuels required for transport, or the energy required to build infrastructure, be it industrial, terrestrial, port or otherwise.

REFERENCES

Bessereau G., Saniere A., Quelles ressources en charbon?, Panorama 2010, IFP Energies Nouvelles, Rueil malmaison, 2010.

BP, BP Energy Outlook 2030, London, January 2012.

CIFOPE, Séminaire sur la Politique Énergétique en Afrique, Lomé, Togo, 1986.

Dargay J., Gately D., Sommer M., Vehicle Ownership and Income Growth, Worldwide: 1960–2030, The Energy Journal, 2007.

Finley M., The Oil Market to 2030 – Implications for Investment and Policy, BP 2012, http://www.bp.com/liveassets/bp_internet/globalbp/globalbp_uk_english/reports_and_publications/statistical_energy_review_2011/STAGING/local_assets/pdf/The_Oil_Market_2030.pdf

Gabriel S., Saniere A., Quelles ressources en combustibles nucléaires?, Panorama 2010, IFP Energies Nouvelles, Rueil malmaison, 2010.

IEA, Energy for all, Financing access for the poor, Paris, 2011.

IEA, UNDP, UNIDO, Energy poverty, How to make modern energy access universal, Paris, 2010.

Kostiuk V.V., Makarov A.A., ed., Énergie et géopolitique, Nakoua, Moscou, 2011, ISBN 978-5-02-037958-9.

Mathieu Y., Un point sur les resources en hydrocarbures, 1 – Les liquides pétroliers, Panorama 2010, IFP Énergies Nouvelles, Rueil malmaison, 2010.
Mathieu Y., Un point sur les resources en hydrocarbures, 2 – Le gaz naturel, Panorama 2010, IFP Energies Nouvelles, Rueil malmaison, 2010.
MDG, Millennium Development Goals Report, United Nations, New York, 2010.
UN, Sustainable energy for all, United Nations, 2012, [Online] Available from: http://www.un.org/en/events/sustainableenergyforall/background.shtml/[Accessed 01/18/2013].
Vially R., Les hydrocarbures non conventionnels: évolution ou révolution, Panorama 2012, IFP Énergies Nouvelles, Rueil malmaison, 2010.

FURTHER READING

Bairoch P., Le Tiers Monde dans l'Impasse, Idées, Gallimard, 1971.
Energy for World Agriculture, FAO, Rome, 1979.
Giraud A., Boy de la Tour X., Géopolitique du pétrole et du gaz, Technip, Paris, 1987.
Grenon M., Réserves et ressources énergétiques (non renouvelables), Énergie Internationale 1989–1990, Economica, Paris 1990.
Lacoste Y., Unité et Diversité du Tiers Monde: des représentations planétaires aux stratégies sur le terrain, La Découverte/Hérodote, Paris, 1984.
L'État du Tiers Monde, La Découverte/CFCF, Paris, 1987.
Martin J.-M., L'Économie Mondiale de l'Énergie, Repères, La Découverte, Paris, 1990.

Chapter 7

Case studies: The energy situation in certain countries

In this chapter we succinctly present the energy situations in eleven different countries, in order to illustrate the contrast between them, as they depend on their respective level of economic development, demographics, natural resource endowments, etc. In this introductory book, these studies only are brief summaries of the problems faced by these countries, especially the larger ones. On the educational level, they can be used as a starting point for further studies or as examples for the analysis of other countries' situations.

The list of countries was selected to cover as wide a range of situations as possible while limiting the number of cases. It includes high-income developed countries (France, the United States, and the United Kingdom), the emerging group known as BRICS (Brazil, Russia, India, China, South Africa), a North African exporter of oil (Algeria), a West African country (Côte d'Ivoire), and an Asian exporter of coal and natural gas (Indonesia).

A section is then dedicated to the European energy policy. It describes the main steps of the construction of a common energy strategy and outlines its main features.

The human development indices (HDI) of the countries studied vary between 0.4 for Côte d'Ivoire and 0.91 for the United States. Their vehicle fleets range from 1 vehicle for 60 people in India to 1 for 1.25 people in the United States. Densities range between 8.3 inhabitants per km^2 in Russia, and 378 inhabitants/km^2 in India, and land area between 244,000 km^2 for the UK and 17 million km^2 for Russia.

Table 7.1 summarizes the key energy indicators in these countries.

The reader may refer to Figure 3.10 in section 3.6.7 for the historical evolution of energy intensities of five of these countries.

General and economic data presented in this chapter are based on databases provided by the World Bank, which are available online (World development index WDI), while the data on primary energy consumption comes from BP statistical review 2012. Data on domestic energy prices comes mainly from Germany's GIZ.

7.1 ALGERIA

With a population of 36 million inhabitants and an area of 2.38 million km^2, resulting in a density of 15.1 inhabitants per km^2, Algeria (Figure 7.1) is the largest country in North Africa.

Table 7.1 Energy indicators.

	PPP GDP $ 2011/capita	*Electricity koe/capita*	*PEC*[1] *(t/capita)*	CO_2 *(MWh/cap.)*	*Vehicles nb/capita*
Algeria	8.37	1.14	3	0.94	0.11
Brazil	11.50	1.36	2	2.17	0.16
China	8.45	1.94	5	2.61	0.05
Côte d'Ivoire	1.73	0.50	0.35	0.20	0.02
France	35.65	3.71	6	7.39	0.57
India	3.62	0.45	1	0.56	0.02
Indonesia	4.53	0.61	2	0.58	0.07
Russia	19.94	4.83	12	6.13	0.27
South Africa	10.79	2.50	9	4.42	0.07
UK	35.94	3.16	8	5.62	0.52
United States	48.89	7.28	18	12.71	0.80

1 Per capita primary energy consumption.

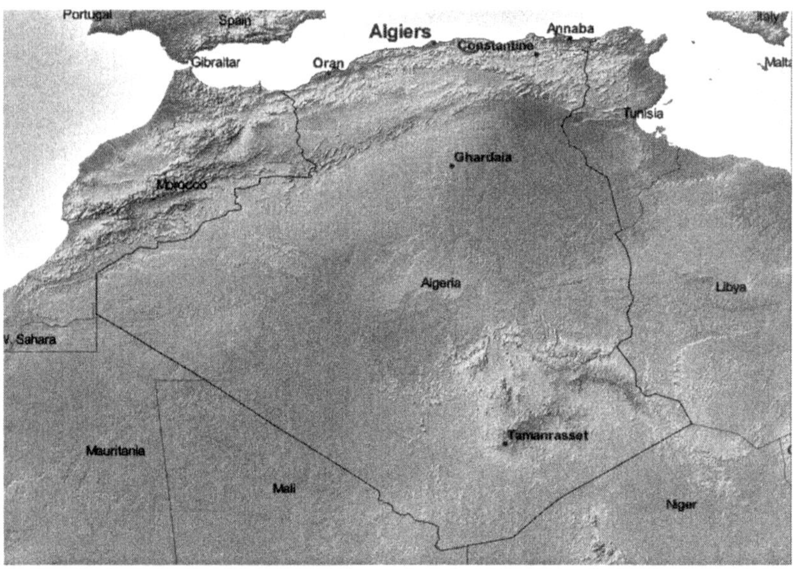

Figure 7.1 Algeria.[1]

Its urbanization rate is 76% and its fleet remains modest (one vehicle per nine inhabitants). Electricity consumption per capita is relatively low (0.94 MWh/capita), although its electrification rate is very high (98%), access to electricity and natural gas being considered a constitutional right.

1 The country maps were built by E. Wey from Natural Earth data. Mapping differs from the usual Mercator cylindrical projection. It uses the EPSG4326 projection of the European Petroleum Survey Group.

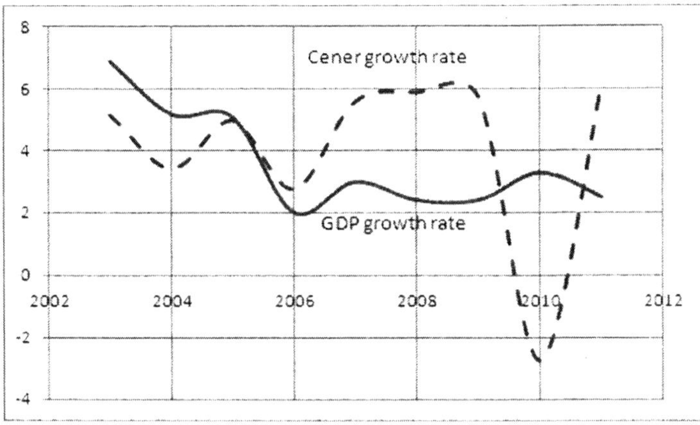

Figure 7.2 Annual growth rate of GDP and primary energy consumption in Algeria.

Its HDI is 0.698, making it a middle-ranking country among those studied in this chapter.

As shown in Figure 7.2, its rate of GDP growth (solid line) varies between 2% and 6%. The growth rate of its energy consumption (dashed curve) is fairly stable, between 4% and 5%, except for 2010.

The price of gasoline and diesel in 2008 were respectively 34 and 20 U.S. cents per liter and that of 1 kWh was around two U.S. cents in 2010, i.e. between three and five times lower than the average of our sample of countries.

With a lot of conventional hydrocarbon resources, Algeria is a member of OPEC that exports nearly three times its own consumption of oil and natural gas, which represents a little more than one third of its GDP. It is the fourth largest oil producer in Africa and the eighth largest natural gas producer. Its reserves/production ratio *R/P* was 19.3 years in 2011 for oil and 57.7 years for natural gas.

Its main customers for oil are North America (mainly the U.S.), which accounts for 40.5% of its exports, Europe (38.5%) and Asia (19.3%).

Its natural gas is exported partly (65%) via pipelines (Medgaz, Transmed, MEG), and partly (35%) in the form of Liquefied Natural Gas (LNG).

The energy intensity of Algeria is relatively low (0.136 koe/$2,011 PPP).

Figure 7.3 shows the structure of the final energy balance in Algeria, hydrocarbons representing about 70% of total energy consumption (27.5 Mtoe in 2007).

The main axes of Algeria's national energy policy are the following:

- developing potential hydrocarbon resources and consequently increasing the national production capacity;
- developing potential alternative energy sources, such as renewable energy, in particular solar energy, of which the country has considerable "reserves", as well as civilian nuclear energy;
- developing the national industry, backed by an efficient infrastructure, to ensure the satisfaction of the domestic demand growth in electricity, natural gas and fuels;

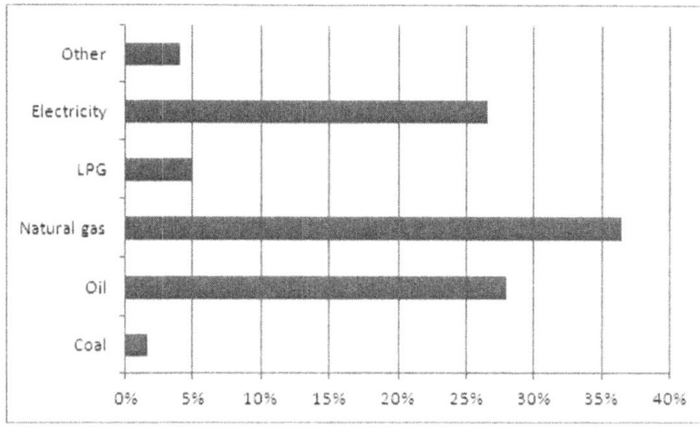

Figure 7.3 Structure of the primary energy balance in Algeria.

- developing the country's gross hydrocarbon resources, including natural gas, which can foster the development of the national industry thanks to petrochemical industries;
- optimizing the exports' value, including through downstream integration on promising markets, particularly for natural gas;
- international deployment to secure new oil and gas reserves in the world;

Sonatrach is the national oil and gas company, with which foreign companies wishing to intervene in Algeria must deal. Created after the country's independence, Sonatrach is now the primary company in the country, accounting for 30% of GDP. The petroleum products market in Algeria is mainly occupied by Naftal, a subsidiary of Sonatrach.

Sonelgaz is the national company responsible for electricity generation and distribution of electricity and gas in Algeria.

In 2002 the Algerian Ministry of Mines established a Committee for regulating electricity and gas (CREG). In 2004, a law on the costs of diversification of the electricity production established preferential tariffs for purchase of electricity from renewable sources (or cogeneration), in order to promote the development of such sources.

7.2 BRAZIL

With a population of 197 million inhabitants, an area of 8,515,000 km^2 and a resulting density of 23.1 inhabitants per km^2, Brazil (Figure 7.4) is the largest economic power in South America.

Brazil has put in place a policy for developing domestic energy for many years, including oil and ethanol, as mentioned section 2.4.4.2.

Important offshore oil resources have been discovered recently. If they are exploited, Brazil could become a major world producer.

Figure 7.4 Brazil.

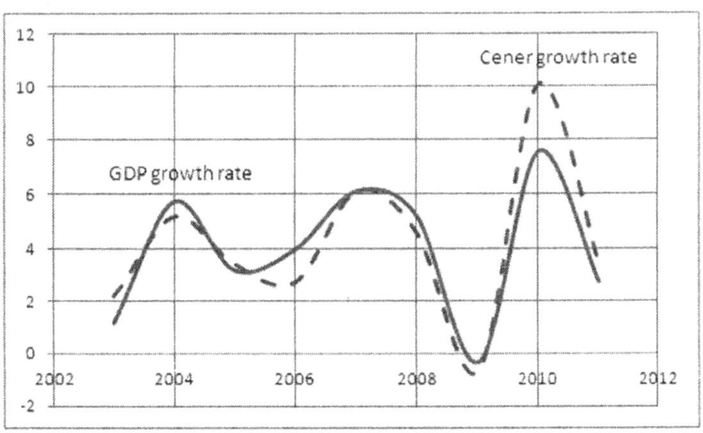

Figure 7.5 Annual growth rate of GDP and primary energy consumption in Brazil.

The country's urbanization rate is 84% while its vehicle fleet remains modest (one vehicle per 6 inhabitants). Per capita electricity consumption is average (2.17 MWh/capita).

Its HDI is 0.813, placing it in the upper range of the countries studied in this chapter.

As shown in Figure 7.5, the rate of GDP growth (solid line) varies between 3% and 6%, except for 2009, because of the impact of the global economic crisis that started in 2008 (an impact that will be noticeable for almost all countries studied in this chapter). The growth rate of energy consumption (dashed curve) is very well correlated.

The prices of gasoline and diesel in 2008 were respectively 126 and 103 U.S. cents per liter and that of 1 kWh 20 U.S. cents in 2010, slightly higher than the average of our sample of countries for fuels, and almost double for electricity.

The production of renewable energy is particularly developed in Brazil, hydropower accounting for 29% of total consumption (270 Mtoe in 2011) and biomass for 21% (Figure 7.6).

In 2011, Brazil was the largest producer of liquid fuels in South America, with 2,700 million barrels/day, 75% oil, thus achieving self-sufficiency for this form of energy.

Its energy independence rate of 90% is the result of deliberate policies pursued by the State for decades.

Brazil's energy strategy is based on the development of its internal resources, particularly in terms of hydropower, uranium, biofuels and agro-energy, fostering a model of rural development.

Projects currently being developed include:

- the development of infrastructure by Electrobras and Gdf Suez;
- the improvement of forest management in the Amazon;
- the Belo Monte dam (production for 26 million households);
- the development of nuclear energy (in the wake of the Angra plant project).

Although the amount of Brazil's per capita CO_2 emissions is relatively low (2 t/capita), its large population makes it the fifth emitter of greenhouse gases, half of the emissions being due to deforestation.

The national oil and gas company PetroBras is the leading operator in the entire chain, from exploration and production to distribution. It held a monopoly until the opening up of the sector to international and other national companies in 1997. The national gas network is managed by PetroBras.

The government agency that oversees the oil operations is the National Petroleum Agency (ANP). ANP has recently taken charge of ethanol production too.

Brazil has encouraged the privatization of its electricity sector since 1996. However, Electrobras remains the main operator in the area, and most of the production

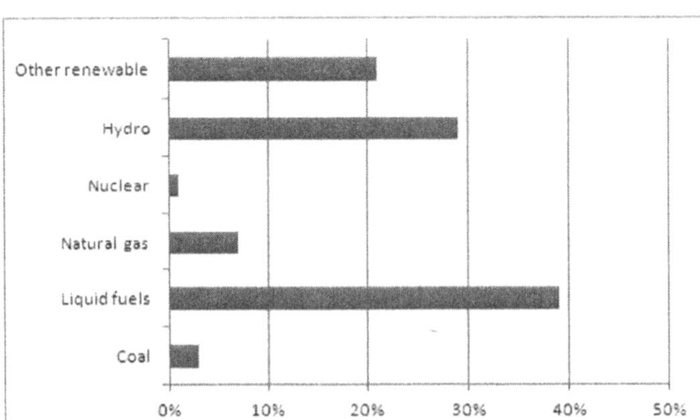

Figure 7.6 Structure of the primary energy balance in Brazil.

is carried out by national companies, while 65% of the distribution is now private. The country's two nuclear power plants are the property of Electronuclear, a subsidiary of Electrobras.

7.3 CHINA

With a population of 1.344 billion inhabitants, an area of 9.6 million km^2, and a density of 140 inhabitants per km^2, China (Figure 7.7) is the second-largest economy in the world.

Its urbanization rate is only 49% and its vehicle fleet, which is rapidly increasing, remains small (1 vehicle per 21.8 inhabitants). Per capita electricity consumption is average (2.61 MWh/capita).

Its HDI is 0.772, placing it in the upper range of the countries studied in this chapter.

As shown in Figure 7.8, the rate of GDP growth (solid line) varied between 10% and 14% during the last ten years, and remains at a high level despite the global economic crisis of 2008 and subsequent years. After a significant decrease between 2004 and 2008, the growth rate of energy consumption (dashed curve) has stabilized around 8%.

The prices of gasoline and diesel were respectively 99 and 101 U.S. cents per liter in 2008 and that of 1 kWh 7 U.S. cents in 2010, a little higher than the average of our sample of countries for fuels, and a little less for electricity.

Over two-thirds of the energy consumption (2613 Mtoe in 2011) is based on coal, a resource China is particularly well endowed with (Figure 7.9). Oil consumption has also increased a lot in recent decades, leading to a great rise in importing oil from the 1990s onwards. It has become the second largest consumer of oil after the United States, and the largest importer.

Its national energy policy's objective is to try and reduce dependence on imports. As the energy prices do not reflect well the costs of supply, China has decided to

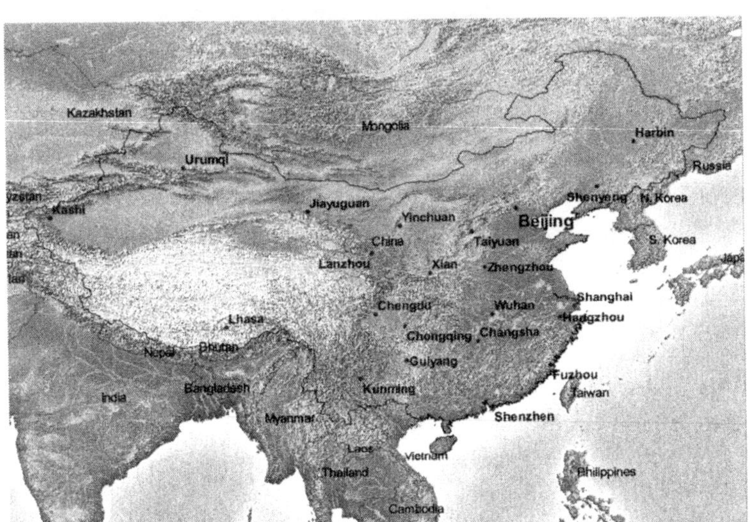

Figure 7.7 China.

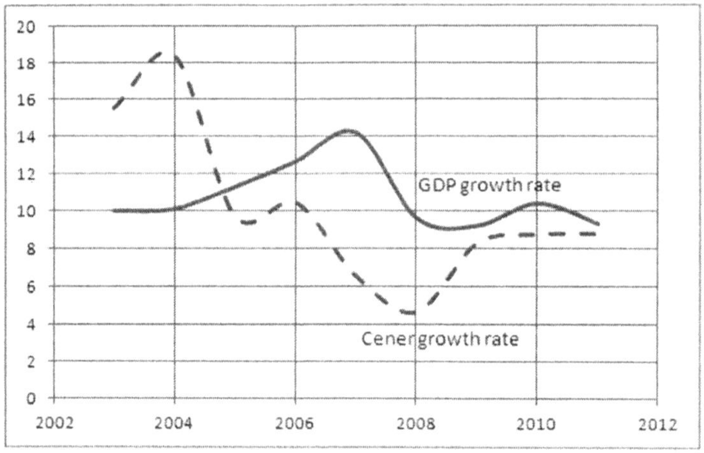

Figure 7.8 Annual growth rate of GDP and primary energy consumption in China.

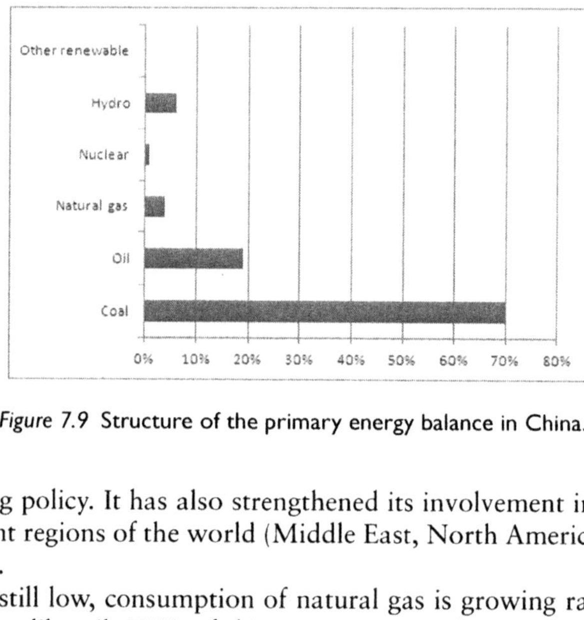

Figure 7.9 Structure of the primary energy balance in China.

change its pricing policy. It has also strengthened its involvement in foreign oil companies in different regions of the world (Middle East, North America, Latin America, Africa and Asia).

Even if it is still low, consumption of natural gas is growing rapidly (by 5% per year). However, unlike oil, 75% of this consumption relies on domestic production, which was in surplus until 2007. About half of the imports are in the form of LNG, the remainder being delivered by pipeline, mainly from Central Asia. China's main LNG suppliers are Australia (30%), Qatar (19%), Indonesia (17%) and Malaysia (13%).

China is the largest producer and consumer of coal, accounting alone for more than half of global consumption. While two decades ago it was exporting coal, it became an importer in 2009.

This explains why, although per capita CO_2 emissions are moderate (5 t/cap), the country is the biggest contributor to the greenhouse effect, before the United States, and why its CO_2 intensity is very high (0.62 t/$ PPP), similar to that of Russia. In addition, 5 Chinese cities are among the 10 most polluted cities in the world.

Energy intensity has declined significantly in recent decades, and continues to do so.

Its electric intensity is among the highest of the studied countries (0.308 MWh/$ PPP), and its consumption is expected to continue growing. Its 2011 installed capacity of 1,073 GW is nearly twice that of 2005 (519 GW). This is why China has decided to develop the production of electricity from natural gas and renewable energy (Figure 7.10).

The National Development and Reform Commission (NDRC) is responsible for defining (among other things) the energy policy, with the assistance of other ministries. In January 2008 the National Energy Administration (NEA) was established to oversee energy operators.

In the 1990s, China restructured its oil and gas companies to form two entities, known as China National Petroleum Corporation (CNPC) and China Petroleum and Chemical Corporation (Sinopec), which dominate the sector as a whole. The former focuses on the entire upstream sector, and the latter on refining and distribution. A third major player is the China National Offshore Oil Corporation (CNOOC), responsible for offshore operations.

Comparatively, the coal industry is much less concentrated, with the 10 largest companies controlling only 30% of the market. Shenhua Coal, the largest company, controls only 10% of it. China National Coal Import and Export Corporation works in partnership with foreign investors in the field.

In 2002, China completed the disintegration of its national electricity company, the State Power Corporation (SPC), which had held a monopoly in the sector, to separate generation, transmission and distribution. Since then, five national companies have produced half of the electricity, the rest being provided by independent producers. The transmission and distribution activities were taken over by two new companies, Southern Power Company and the State Power Grid Company, the latter operating in the north of the country and the former in the south.

In 2002, the State Electricity Regulatory Commission (SERC) was established to oversee the power sector and reduce power outages.

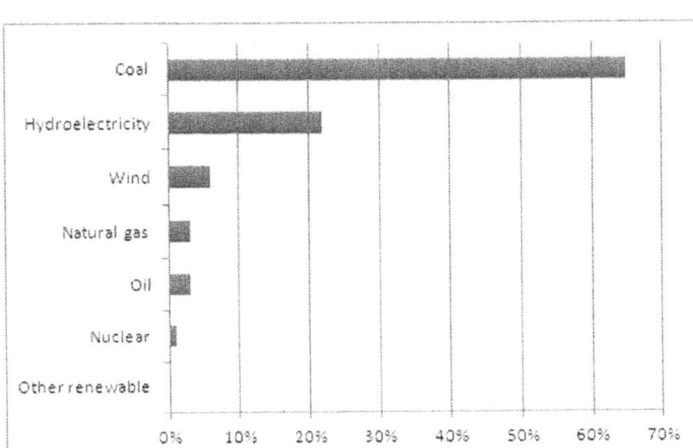

Figure 7.10 Structure of electricity production in China.

7.4 CÔTE D'IVOIRE

Côte d'Ivoire (Figure 7.11) is a country in Western Africa with a population of 20 million inhabitants, an area of 322,000 km^2, and a density of 62.5 inhabitants per km^2.

Its urbanization rate is 51% and its vehicle fleet is very small (1 vehicle per 50 inhabitants). Per capita electricity consumption is very low (200 kWh/capita).

Its HDI is 0.4, the lowest of the countries surveyed in this chapter.

As shown in Figure 7.12, the rate of its GDP growth, which was positive with an upward trend, dropped after 2009 due to the global economic crisis, whose consequences for the Ivorian economy were increased by the serious internal political crisis

Figure 7.11 Côte d'Ivoire.

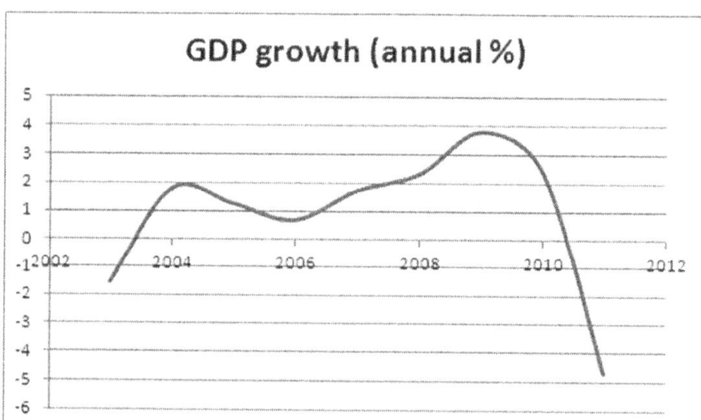

Figure 7.12 Annual growth rate of GDP in Côte d'Ivoire.

which arose out of a dispute between the former president Laurent Gbagbo and the newly elected president, Alassane Ouattara. This crisis was resolved in April 2011.

The prices of gasoline and diesel in 2008 were respectively 133 and 120 U.S. cents per liter and that of 1 kWh, 10 cents in 2010, significantly higher than the average of our sample of countries for fuels and equal to that value for electricity. The cost of mobility thus is particularly high in Côte d'Ivoire.

Despite the recent development of services, the Ivorian economy is dominated by agriculture. Côte d'Ivoire is characterized by a variety of conventional energy resources such as oil and gas. The use of renewable energy also plays a fairly important role (Figure 7.13).

Energy is mainly consumed by its residential sector, which represents almost two thirds of the total (10 Mtoe in 2011). The remainder is roughly equally divided between industry, commerce and transport.

Energy policy in Côte d'Ivoire is primarily aimed at enhancing national energy resources such as oil and gas, hydro energy, biomass and solar energy. It aims at meeting the needs at the lowest possible cost and at entirely covering the national territory in order to impede the rural exodus by creating on-site living conditions that are relatively comfortable for all.

In addition to developing its oil and natural gas exports, in 1993 Côte d'Ivoire adopted a "butanization" policy to combat deforestation. Marketed in two forms, conditioned butane and bulk butane, butane gas then became a secondary activity of the national oil company PETROCI.

In 2005, rural electrification programs undertaken by the Ivorian government had helped to electrify 2,413 localities among a total of 8,513, as well as to achieve a national coverage rate of 28.3% and a population electricity access rate of 70.9%. More than two out of three inhabitants have access to electricity and only 29% of the population resides in non-electrified communities.

Since 1975, the National Petroleum Operations Company (PETROCI or *Société Nationale d'Opérations Pétrolières* in French) has been in charge of promoting the Ivorian sedimentary basin and valorizing its oil and gas resources by developing exploration of oil and gas in the country.

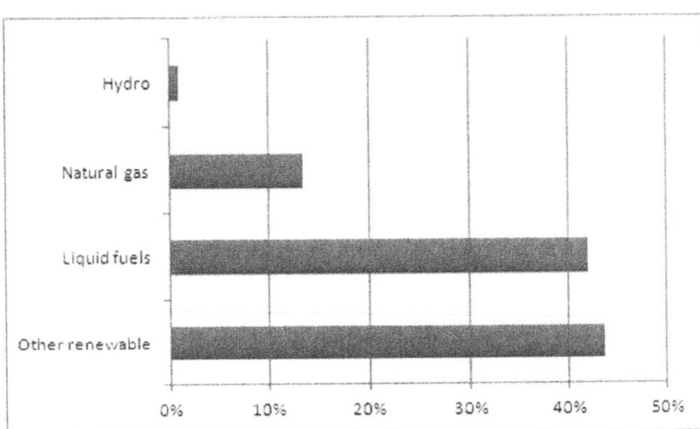

Figure 7.13 Structure of primary energy balance in Côte d'Ivoire.

The Ivorian Electricity Company (CIE or *Compagnie Ivoirienne d'Electricité* in French, created on August 24, 1990) is responsible for the operation and maintenance of the generation facilities, the transmission and distribution networks of electricity owned by the State, as well as the management of the client portfolio. The Ivorian government also put in place three companies in the electricity sector in 1998:

- SOGEPE (*Société de Gestion du Patrimoine du secteur de l'Électricité* in French) responsible for asset management and tracking of flow of finances;
- ANARE (National Regulatory Authority for Electricity, *Autorité Nationale de Régulation de l'Électricité* in French) responsible for supervising the operator and independent producers, responsible for arbitration of disputes between different actors and clients;
- SOPIE (*Société d'Opération Ivoirienne d'Électricité* in French) responsible for monitoring the management of energy flows between producers and consumers, as well as the management of investment projects.

7.5 FRANCE

With a population of 65 million inhabitants, an area of 555,000 km^2, and a density of 119.2 inhabitants per km^2, France (Figure 7.14) is one of the largest economies in Europe.

France has very few energy resources, following the closure of almost all of its coal mines, which had ensured its economic development in the nineteenth century and the first half of the twentieth century.

Compared to other industrialized countries in the EU or the OECD, the French energy situation mainly distinguishes itself by its very low level of national resources.

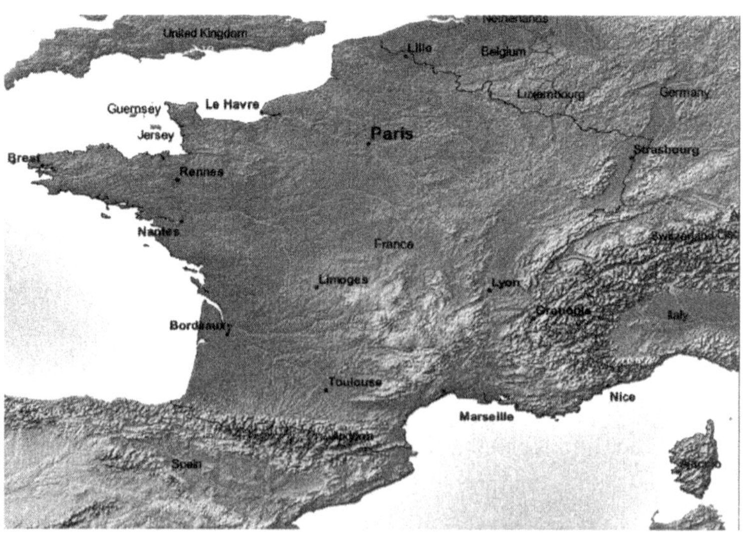

Figure 7.14 France.

This situation has had a significant impact on France's energy policy, and in particular on its choice to develop nuclear plants since 1974.

Regarding the final energy consumption, by contrast, the situation in France is not very different from that of its EU partners.

Its urbanization rate is 85% and its fleet of vehicles rather large (1 vehicle per approximately 1.76 people). Its per capita electricity consumption is high (7.39 MWh/capita).

Its HDI is 0.884, one of the highest of all the countries surveyed in this chapter.

As shown in Figure 7.15, its rate of GDP growth (solid line) is around 2%, with the exception of 2009 because of the impact of the global economic crisis. The growth rate of energy consumption (dashed curve) is fairly well correlated.

The prices of gasoline and diesel in 2008 were respectively 152 and 145 U.S. cents per liter and that of 1 kWh 17 cents in 2010, that is to say nearly 50% higher than the average of our sample of countries.

The share of nuclear energy in the primary energy consumption (243 Mtoe in 2011) is particularly high in France, followed by oil and natural gas (Figure 7.16).

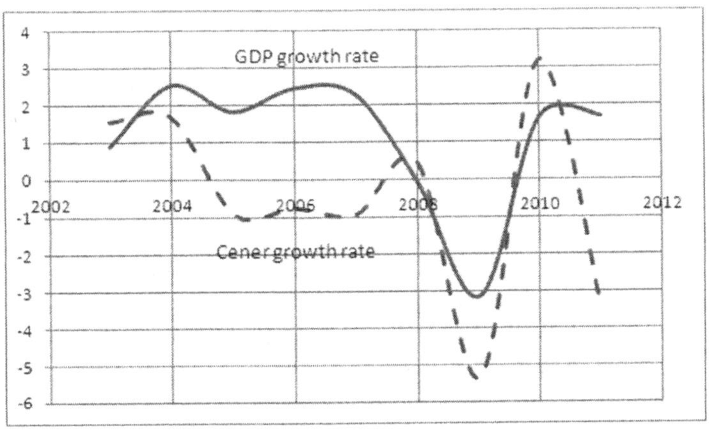

Figure 7.15 Annual growth rate of GDP and primary energy consumption in France.

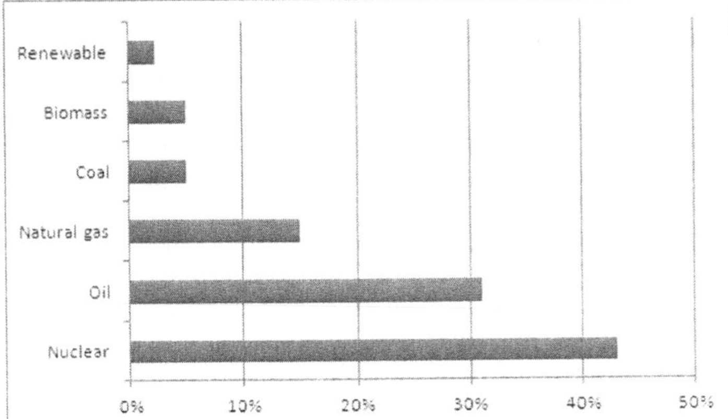

Figure 7.16 Structure of the primary energy balance in France.

The amount of CO_2 emissions is lower in France (6 t/capita) than in other EU countries, due to the production of electricity from nuclear origin.

The French energy strategy is based on four principles:

- security of supply;
- economic competitiveness;
- sustainable development;
- equitable access to energy for all its citizens irrespective of their place of residence.

At the institutional level, recent decades have been marked by the deregulation of the energy sector, with the opening up to competition of gas and electricity markets, formerly Gaz de France (GDF) and Électricité de France (EDF) monopolies, in the wake of the deregulation process in place in the United States and Europe.

Its main oil and gas companies are Total and GDF Suez.

Approximately 11 million consumers are connected to the natural gas distribution networks. These customers are supplied by 25 natural gas Distribution System Operators (DSOs) of very diverse sizes, supervised by the Commission for Energy Regulation (*Commission de Régulation de l'Énergie* in French). The most important is GRDF (*Gaz Réseau Distribution France*, a subsidiary of GDF SUEZ distribution), which distributes more than 96% of the market;

For electricity, EDF is by far the largest producer, transporter, distributor and supplier of electricity. The public transport network electricity is operated by RTE (*Réseau de Transport de l'Électricité*).

The public distribution networks are owned by municipalities, which may entrust the management to ERDF (95% of the distribution of continental metropolitan area), or Local Distribution Companies (LDCs) through concession contracts.

Electricity transmission and distribution managers exercise monopolies regulated by the Commission of Energy Regulation.

7.6 INDIA

With a population of 1.24 billion people and an area of 3.3 million km^2, and a density of 377.7 inhabitants per km^2, India (Figure 7.17) is one of the largest countries in Asia. Given the rate of its population growth, it will soon become the most populous country in the world.

Its urbanization rate is very low (31%) and its fleet of vehicles small (1 vehicle per 20.55 inhabitants). Although its per capita electricity consumption (0.56 MWh/capita) is low, the overall electrification rate is 66%, and 50% in rural areas.

Its HDI is 0.519, placing it in the lower range of the countries studied in this chapter.

As shown in Figure 7.18, its rate of GDP growth (solid line) has stayed except for 2009, between 8% and 10%. The growth rate of its energy consumption (dashed curve) is quite stable, ranging between 6% and 8%.

The prices of gasoline and diesel in 2008 were respectively 109 and 70 cents per liter and that of 1 kWh 8 U.S. cents in 2010, slightly higher than the average of our sample of countries for gasoline and electricity and a little lower for diesel.

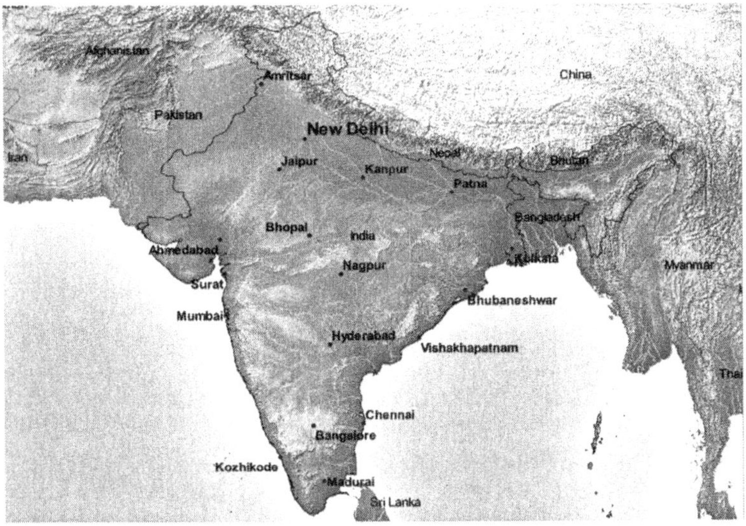

Figure 7.17 India.

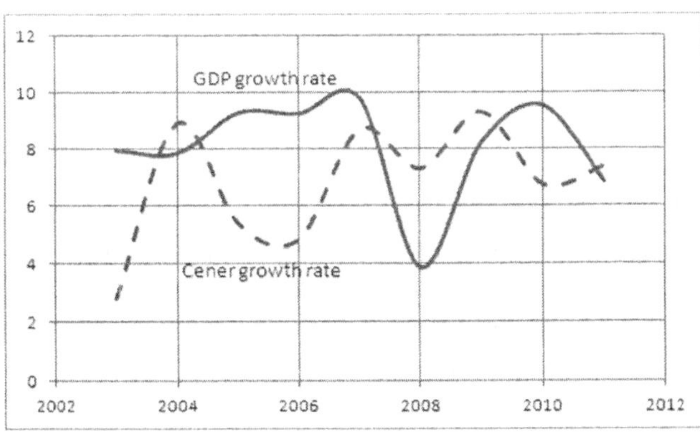

Figure 7.18 Annual growth rate of GDP and primary energy consumption in India.

Despite its very low per capita energy consumption (0.45 koe/capita), it is the fourth largest consumer in the world after the U.S., China and Russia, due to its population size. Its oil and natural gas imports are growing.

Coal and oil represent about two-thirds of its total consumption (559 Mtoe in 2011), and renewable energy (biomass, waste) approximately one quarter (Figure 7.19).

Although India has oil resources, its consumption is four times greater than its production. The country's situation regarding natural gas and coal is better as its national production represents three quarters of its consumption.

80% of its electricity generation is provided by thermal power plants (56% from coal), 15% by hydropower, and the remainder by nuclear power and renewable energy. The power plant installed capacity is about 180 GW.

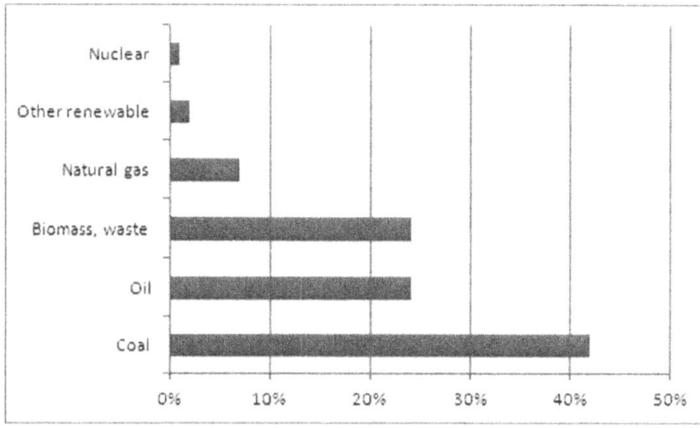

Figure 7.19 Structure of the primary energy balance in India.

The main oil and gas upstream operator is the national Oil and Natural Gas Corporation (ONGC) which provided nearly three quarters of the oil production in 2009–2010. Nevertheless, the private sector is playing an increasingly significant role. In terms of refining and distribution of petroleum, the national company Indian Oil Corporation (IOC) is the largest.

The gas price regulator is the Administered Pricing Mechanism (APM), and the transport and distribution of gas are the monopoly of the Gas Authority of India Ltd. (GAIL).

India has a chronic shortage of installed electricity capacity of about 10%, and 15% as regards production, with supply failing to meet demand growth. The great blackout of July 2012 reported on in section 2.5.2 was a consequence of this situation. India is thus implementing incentives for large consumers to reduce their demand for electricity.

7.7 INDONESIA

With a population of 242 million inhabitants and an area of 1.9 million km^2, and a density of 127.2 inhabitants per km^2, Indonesia (Figure 7.20) is the largest archipelago in the world, comprising more than 17,000 islands, among which only 6,000 are inhabited.

The four largest islands are Java, Sumatra, Kalimantan (Indonesian part of Borneo) and New Guinea (shared with Papua New Guinea).

The country's main industries are oil and natural gas, textiles and clothing, as well as mines. The main agricultural products are palm oil, rice, tea, coffee, spices and rubber.

Its urbanization rate is average (50%), while its fleet (1 vehicle per 18.2 inhabitants) and its per capita electricity consumption (0.58 MWh/capita) are low.

Its HDI is 0.726, placing it among the middle-ranking countries studied in this chapter.

Figure 7.20 Indonesia.

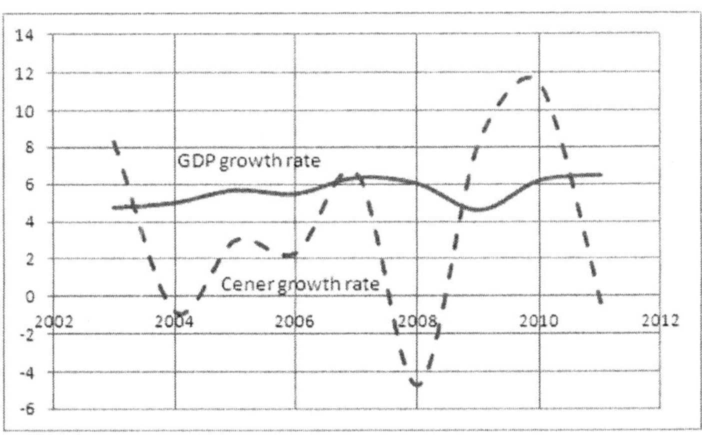

Figure 7.21 Annual growth rate of GDP and primary energy consumption in Indonesia.

As shown in Figure 7.21, its GDP growth rate (solid line) ranged between 4% and 6% during the last ten years. The growth rate of energy consumption (dashed curve) varies a lot, oscillating between –5% and 12%.

The prices of gasoline and diesel in Indonesia were respectively 50 and 42 U.S. cents per liter in 2008 and that of 1 kWh 6 cents in 2010, i.e. nearly half of those in our sample of countries.

With its many fossil fuel resources, Indonesia has stopped exporting oil since 2004, but still exports large quantities of coal (making it the second largest coal exporter in the world) and natural gas (6th in the world). The distribution of its primary energy consumption (148 Mtoe in 2011) is given in Figure 7.22.

86% of its electricity generation is provided by thermal power plants (40% from coal), 8% from hydropower, with the rest being supplied by nuclear power and renewable energy sources, including geothermal. The power plant installed capacity is about 28 GW.

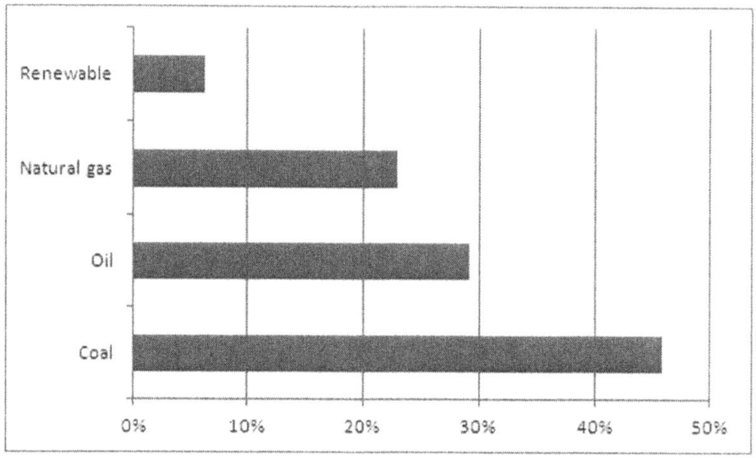

Figure 7.22 Structure of the primary energy balance in Indonesia.

At the institutional level, the Indonesian Ministry of Energy and Mineral Resources is responsible for the oil and gas sectors, supervision being provided by BP Mipgas. The national company PT Petermina only accounts for 15% of crude oil production, the rest being provided by foreign companies.

The main electricity company, PT PLN (Perusahaan Listrik Negara), is state-owned. It provides 85% of the domestic production. The electrification rate is only 65%, and supply disruptions are frequent. To develop this sector, the Indonesian government expects increased use of coal.

7.8 RUSSIA

With a population of 142 million inhabitants and an area of 17.1 million km^2, with a density of 8.3 inhabitants per km^2, Russia (Figure 7.23) is one of the major world powers.

Its urbanization rate is 74% and its fleet is rather developed (1 vehicle per 3.71 people). Per capita electricity consumption is high (6.13 MWh/capita).

Its HDI is 0.817, placing it in the upper range of the countries studied in this chapter.

As shown in Figure 7.24, its rate of GDP growth (solid line) was between 6% and 9% before the global economic crisis of 2008 and subsequent years. The growth rate of its energy consumption (dashed curve) is well correlated.

The prices of gasoline and diesel in 2008 were respectively 89 and 86 U.S. cents per liter and that of 1 kWh 8 cents in 2010, that is to say nearly 20% less than the average for our sample of countries.

Russia's hydrocarbon reserves are considerable, making it one of the largest producer and exporter of oil and natural gas.

Its oil production exceeds 10 million barrels/day and ranks second in the world, after Saudi Arabia. 70% of this oil is exported. Its *R/P* ratio is 23.5 years. Figures 6.7 and 6.8 show Russia's oil and gas balances.

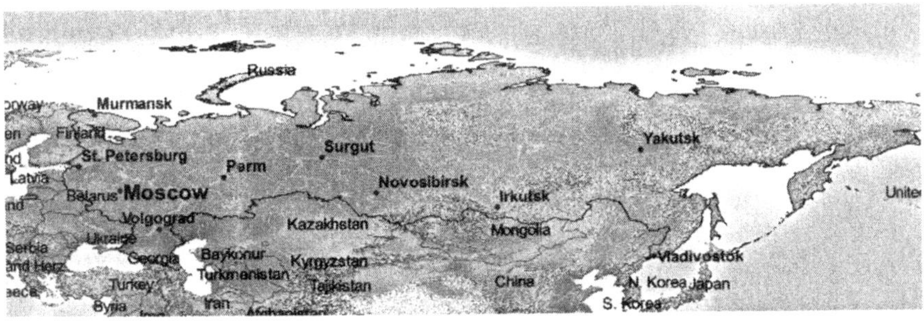

Figure 7.23 Russia.

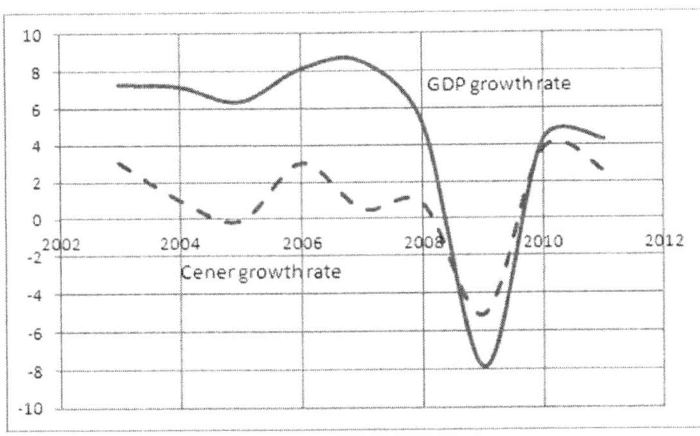

Figure 7.24 Annual growth rate of GDP and primary energy consumption in Russia.

Russia holds the largest natural gas reserves in the world (21.4% of total). Its output is 607 billion m^3, or 546 Mtoe, with a *R/P* ratio of 73.5 years.

Although its coal reserves are very significant and rank second behind those of the United States, its production and consumption are relatively limited. The Russian government has recently restructured the coal industry to promote production by independent operators, in order to increase production and to substitute coal for natural gas in electricity production, preferring to reserve gas for exports.

The primary energy consumption in Russia was 686 Mtoe in 2011, with approximately 54% being supplied by natural gas (Figure 7.25).

Although steadily declining since 1995, its energy intensity (0.242 koe/$ PPP) and CO_2 intensity (0.6 t/$ PPP) are quite high. This is only partly explained by the country's climate. Indeed, its production system was inherited from the former Soviet Union, with heavy industries that are still highly energy-intensive and whose energy efficiency remains low.

Two-thirds of Russia's electricity generation is provided by thermal plants, 17% by hydropower, with the rest being supplied by nuclear plants (Figure 7.26).

At the institutional level, Russia has undergone very significant changes over the last two decades, following the collapse of the Soviet Union.

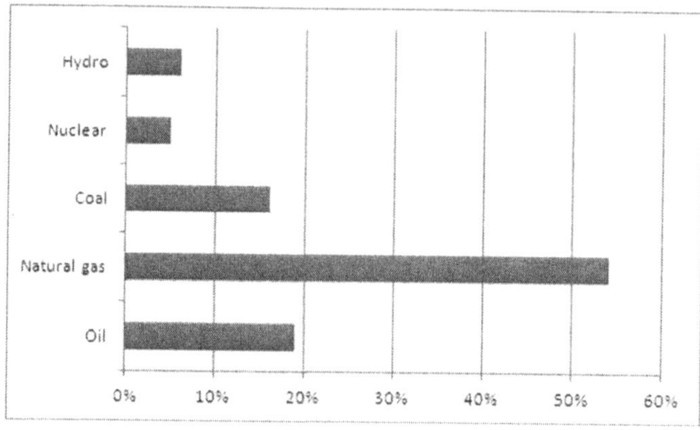

Figure 7.25 Structure of the primary energy balance in Russia.

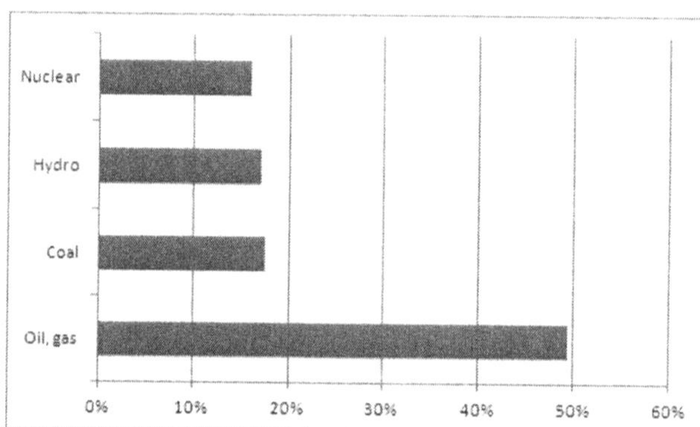

Figure 7.26 Structure of the electricity production in Russia.

Relevant energy ministries in Russia are the Ministry of Natural Resources (permit to operate), the Ministry of Finance (pricing), the Ministry of Economic Development and the Ministry of Energy, which oversees energy policy. For nuclear power, the competent authority is Rosatom, the Russian Federal Agency for Atomic Energy.

The two largest oil companies are Rosneft and Lukoil. The Russian company Transneft holds the monopoly for oil pipeline transportation (80% of exports).

The national company Gazprom provides 80% of the natural gas production and has direct control over 65% of reserves. It also controls the gas pipeline network.

Wholesale and retail pricing is done by regulators such as the Federal Power Commission and the Regional Energy Commissions.

Following the privatization independent producers are now responsible for more than 80% of coal production.

The electricity sector is divided into eight regions, seven of which are interconnected. The Federal Network Company, in which the State has a controlling share of more than 70%, manages almost all the transmission and distribution.

Electricity production was widely privatized during the last decade, with the exception of hydroelectric and nuclear power.

7.9 SOUTH AFRICA

With a population of 51 million inhabitants and an area of 1.2 million km², resulting in a density of 41.5 inhabitants per km², South Africa (Figure 7.27) is the largest economic power of Africa.

On the political front, the major event that marked the country in recent years was the end of Apartheid between 1991 and 1994. Liberalism has since prevailed as an economic model, but it has been tempered by government intervention to reduce the social differences inherited from the past.

Despite the existence of "townships" where many poor people are concentrated around big cities, South Africa's urbanization rate is only 62% and its vehicle fleet remains small (1 vehicle per approx. 15 people). Per capita electricity consumption is relatively high (4.42 MWh/cap) as the electrification rate is around 75%, although this figure drops to only 55% for the rural population. Growth in electricity demand is such that supply is struggling to keep up, resulting in frequent power cuts.

South Africa's HDI is 0.619, which places it among the middle-ranking countries studied in this chapter.

As shown in Figure 7.28, its rate of GDP growth (solid line) varies between 3% and 5%, except for 2009, which relates to the global economic crisis. The growth rate of its energy consumption (dashed curve) is fairly well correlated, except for the year 2005.

The prices of gasoline and diesel in 2008 were respectively 87 and 95 U.S. cents per liter and that of 1 kWh 13 cents in 2010, slightly lower than the average of our sample of countries for fuels, and a little more for electricity.

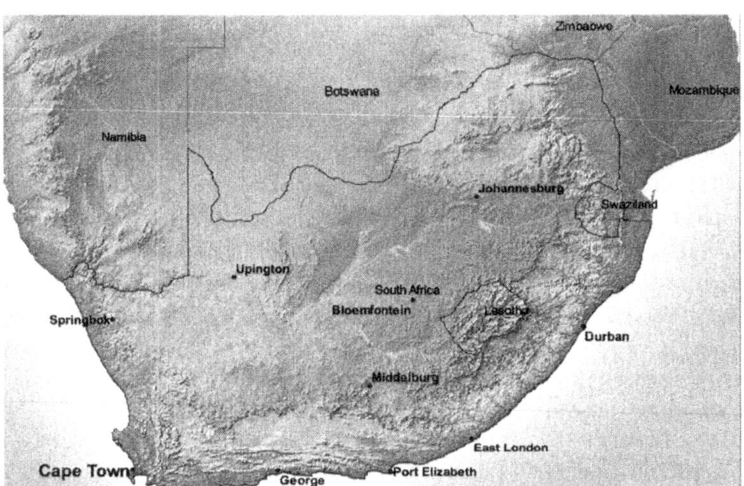

Figure 7.27 South Africa.

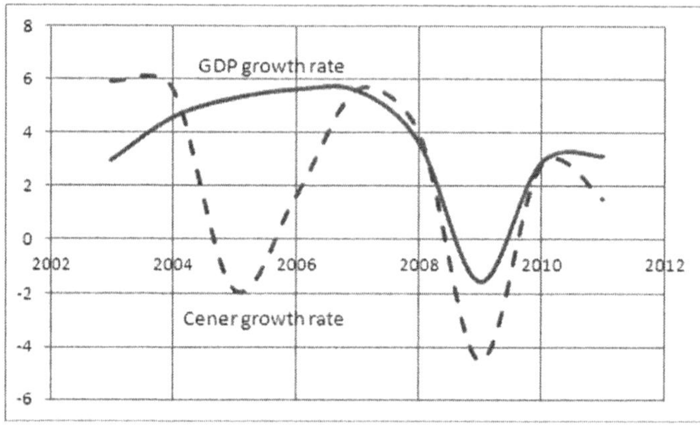

Figure 7.28 Annual growth rate of GDP and primary energy consumption in South Africa.

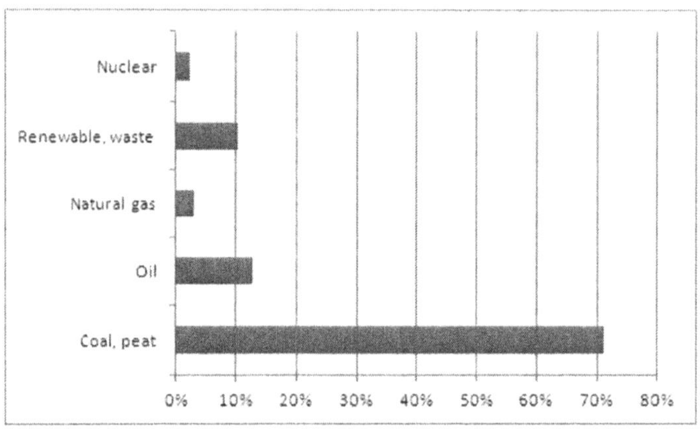

Figure 7.29 Structure of the primary energy balance in South Africa.

With few conventional hydrocarbon resources, South Africa imports two thirds of its consumption of oil and natural gas. Its refining capacity is one of the largest in Africa.

It does have a lot of coal however, which is used for export and for its own needs, and which explains its high CO_2 emissions (9 t/capita). 90% of South Africa's electricity is generated from coal, which is also the primary raw material used for the production of synthetic hydrocarbons, in which the country has been heavily involved for many years.

It could eventually produce large quantities of shale gas, despite a moratorium in 2011 for environmental reasons.

Its primary energy consumption was 126.3 Mtoe in 2011. Energy intensity is relatively low (0.231 koe/$2011 PPP) in South Africa, and only 5.1% of its electricity presently comes from nuclear power.

At the institutional level, in 2005 South Africa created an agency in charge of the national energy policy: the National Energy Regulator of South Africa (NERSA). Its national oil and gas company is called Petroleum Oil and Gas Corporation of South Africa (PetroSA). It operates within South Africa as well as internationally. The Sasol petrochemical company is one of the major operators in the energy sector, given its involvement in the production facilities of liquid fuels from coal or natural gas.

Eskom, the largest electricity company, produces 95% of the country's electricity with an installed capacity of about 50 GW, 85% of which comes from coal. It is also responsible for the distribution network. NERSA encourages independent producers, in particular to promote technologies for off-grid rural areas.

7.10 UNITED KINGDOM

With a population of 63 million inhabitants, an area of 244,000 km², and a density of 257.1 inhabitants per km², the United Kingdom (Figure 7.30) is one of the largest European economies.

Its urbanization rate is 80% and its fleet is significant (one vehicle for every two inhabitants). Per capita electricity consumption is high (5.62 MWh/capita) in the UK.

Its HDI is 0.849, placing it in the upper range of the countries studied in this chapter.

As shown in Figure 7.31, the growth rate of its GDP (solid line) varied between 1% and 3% during the last ten years, with the exception of 2009 because of the impact of the global economic crisis that started in 2008. The growth rate of its energy consumption (dashed curve) is correlated.

The prices of gasoline and diesel in 2008 were respectively 144 and 165 U.S. cents per liter and that of 1 kWh 13 cents in 2010, i.e. 50% higher than the average of our sample of countries for fuels, and 30% higher for electricity.

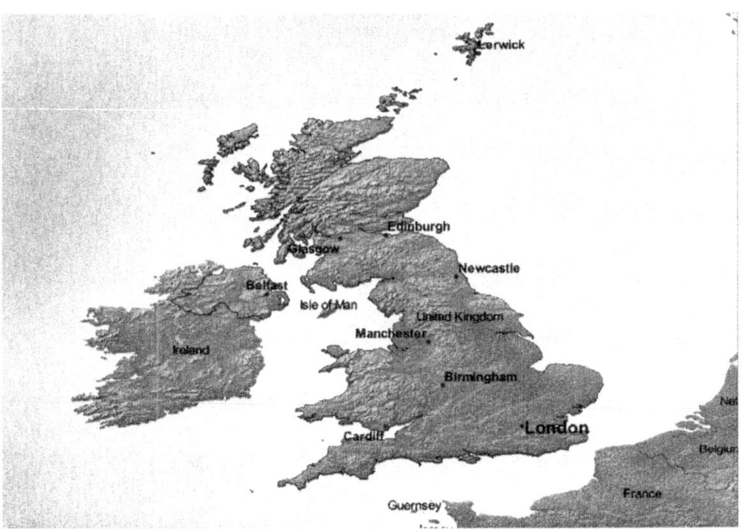

Figure 7.30 United Kingdom.

While deposits in the North Sea have allowed the UK to become an exporter of oil and natural gas, the production has started to decline since the beginning of the century (Figure 8.10) while the consumption of oil has remained stable. This has resulted in the United Kingdom becoming an importer in 2004–2005.

To compensate for this inevitable decline, the UK invests in recovery techniques for oil and gas fields as well as in energy conservation and renewable energy development.

The latter part of its energy policy should lead to a reduction in the amount of CO_2 emissions (8 t/capita).

Coal production, which had dropped significantly over the past twenty years, offset by significantly cheaper coal imports, has now stabilized.

Figure 7.32 shows the structure of the UK primary energy balance (198 Mtoe in 2011).

A more detailed presentation was given in Figure 3.5 presenting a graph for energy flow in 2011.

80% of electricity generation is provided by thermal power plants, with the rest being supplied by nuclear power plants (13%), renewable energy (6%) and hydro (1%). The power plant installed capacity is about 86 GW.

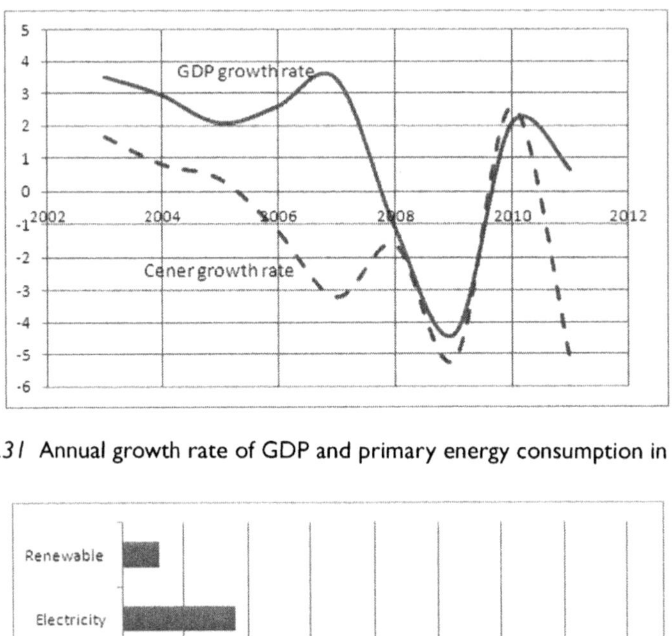

Figure 7.31 Annual growth rate of GDP and primary energy consumption in the UK.

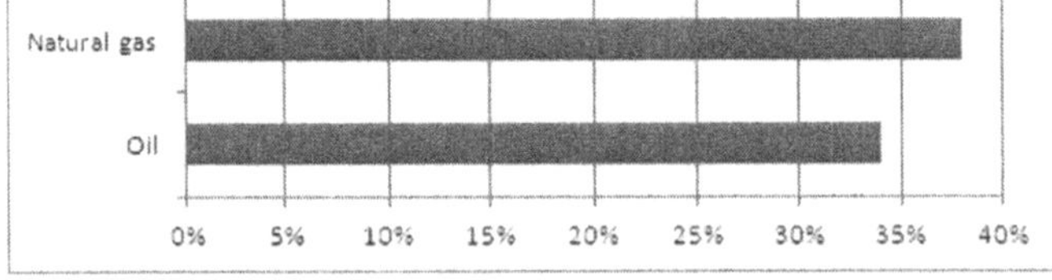

Figure 7.32 Structure of the primary energy balance in the UK.

The main oil company is British Petroleum, BP. The natural gas industry, upstream and downstream, is dominated by private companies, the most important being Centrica, an offshoot of the former British Gas.

Production and distribution of electricity have also been privatized, the largest producer being EDF Energy, which controls nearly all of the nuclear production.

7.11 UNITED STATES

With a population of 312 million inhabitants, an area of 9.8 million km^2, and a density of 31.7 inhabitants per km^2, the United States (Figure 7.33) is the largest economy in the world.

Economic indicators and energy per capita are very high (see Table 7.2). However, their values are progressively moving downward, due to advances in energy conservation (–1.7% per year in energy intensity from 1990 to 2011).

The United States' urbanization rate is 82% and its vehicle fleet considerable (1 vehicle per 1.25 residents). The country's HDI is 0.91, the highest of the countries surveyed in this chapter.

As shown in Figure 7.34, its GDP growth rate (solid line) varied between 2% and 3% during the last ten years, except for 2008 and subsequent global economic crisis years. The growth rate of energy consumption (dashed curve) is fairly well correlated.

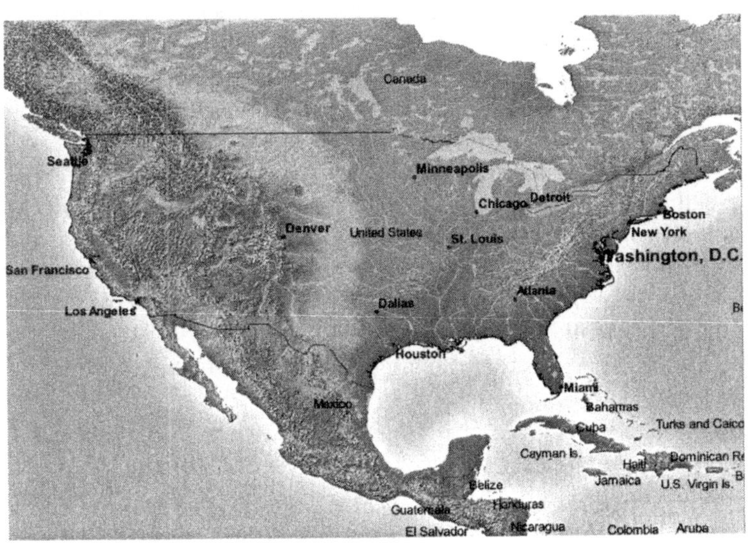

Figure 7.33 United States.

Table 7.2 Energy indicators of the United States.

	PPP GDP $2011/capita	Electricity koe/capita	Cener (t/capita)	CO_2 (MWh/cap.)	Vehicles nb/capita
United States	48.89	7.28	18	12.71	0.80

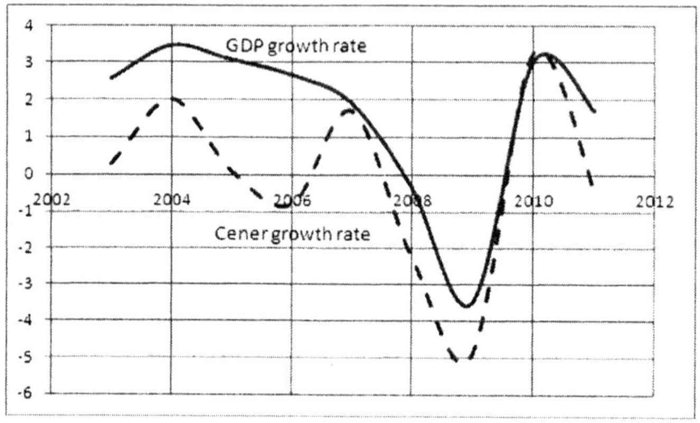

Figure 7.34 Annual growth rate of GDP and primary energy consumption in the US.

The prices of gasoline and diesel in 2008 were respectively 56 and 78 cents per liter and that of 1 kWh 11 U.S. cents in 2010, much less than the average for our sample of countries for fuels and equal for electricity.

As shown in Figure 7.35, 83% of primary energy (2269 Mtoe in 2011) is provided by fossil fuels, which explains the very high CO_2 emissions in the United States. Like China, the country refused to sign the Kyoto Protocol, considering that its implementation could slow its economic growth.

After decades during which its consumption of oil and gas grew faster than its production thereof, the United States has begun to reverse this trend in recent years, even becoming an exporter of natural gas in 2010, thanks to the exploitation of shale gas.

The overall energy dependence of the US, which rose to a peak of 30% in 2005, fell to 19% in 2011 and should be limited to 10% in 2035. Concerning liquid fuels, these figures are 60% in 2005, 45% in 2011 and 37% in 2035. Figure 6.5 shows the oil balance of the United States.

Fossil fuels account for two-thirds of electricity generation, as shown in Figure 7.36, and this is expected to change little in the future, with substantially less coal and more natural gas.

CO_2 emissions, which fell by nearly 10% between 2005 and 2011, are set to grow again in the coming decades, according to forecasts by the U.S. Energy Information Administration, and despite the targets set by the Obama administration, which pledged to reduce them by 17% by 2020.

At the institutional level, the Department of Energy of the United States federal government (DoE) has been responsible for energy policy and nuclear safety since 1973.

The Energy Information Administration (EIA), an independent agency, publishes official statistics on energy on behalf of the government of the United States. EIA collects, analyzes and publishes this data in order to ensure greater market efficiency, to inform policy makers and to improve the public's understanding of energy issues.

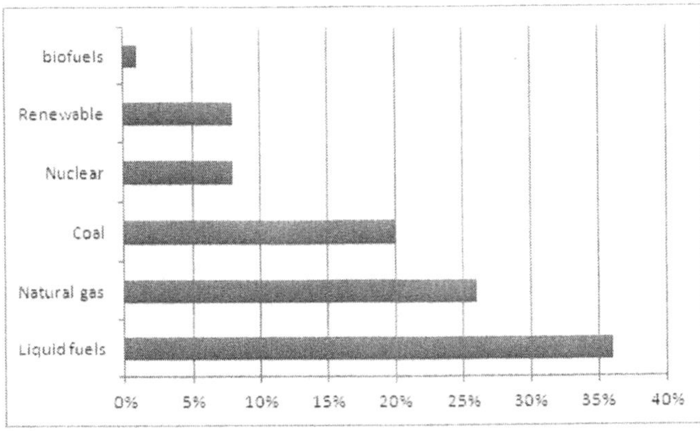

Figure 7.35 Structure of the primary energy balance in the United States.

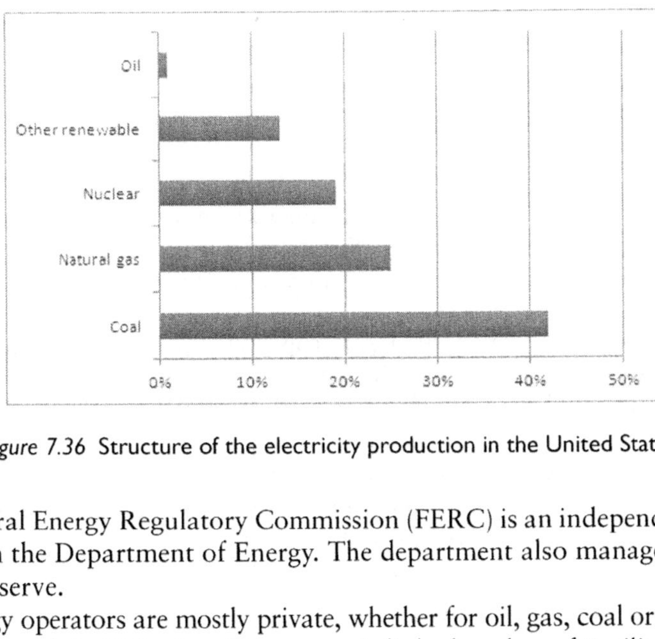

Figure 7.36 Structure of the electricity production in the United States.

The Federal Energy Regulatory Commission (FERC) is an independent regulatory agency within the Department of Energy. The department also manages the Strategic Petroleum Reserve.

The energy operators are mostly private, whether for oil, gas, coal or nuclear energy.

The history of the U.S. electricity sector is linked to that of "utilities", which are vertically integrated companies specialized in the production, transportation and sale of electricity. Their activities are managed and controlled at the State level by regulatory bodies (Public Utility Commission and FERC).

In 1998, 239 private utilities produced 68% of electricity and accounted for 75% of sales, while public utilities in 2009 produced 9% of electricity and accounted for 15% of sales.

The electricity sector has been marked in recent decades by a process of deregulation implemented in three steps:

- PURPA in 1978 (Public Utility Regulatory Act), which amends the legislation applying to utilities, forcing them to purchase electricity from Independent

Power Producers which qualify (Qualifying Facilities) at the utility's avoided cost[2];

- the Energy Policy Act (EPACT) of 1992, further opening competition in the field of production by creating a new category of producers, Exempt Wholesale Generators (EWGs), which, like the Qualifying Facilities, do not fall under the Public Utility Holding Company Act of 1935 (PUHCA);
- finally, the process of opening up supply contracts initiated in many states in the mid-1990s.

This resulted in the creation of a market for independent production (not provided by utilities) and promoted the development of Independent Power Producers (IPP), which would soon take over a significant market share (9% in 1992, 11.2% in 1998, 14% in 1999, 34% in 2002).

7.12 ENERGY IN EUROPE

As shown by the analyses in Chapter 6, the European Union is the largest importer of oil in the world, with 600 Mt/year, or 12 Mb/d. It imported 84.3% of its oil consumption in 2008, and its dependence should increase even further in the future. The situation is similar for natural gas, the share of imports being 62.3% in 2008 and ever increasing.

Given its economic weight (it is the world's largest economy with a GDP equal to 25% of GWP) and its share in world energy imports, it seemed necessary to supplement the previous country studies by a section devoted to the European energy policy, whose pace has finally picked up after the signing of the Treaty of Lisbon.

We already presented data related to Europe in the previous chapters, particularly in Section 5.5, where Table 5.2 gives the values of energy prices for end users in some EU countries, and in Chapter 6, where Figure 6.6 shows the changes in oil production, consumption, imports and exports.

Any reader interested in detailed analyses of energy situations in different European countries will easily find detailed information on the European Commission's website, given in reference, as well as in [EC, 2011].

7.12.1 Evolution through the years

The European construction is the result of a multi-decade process originally initiated by the Treaty of Rome in 1957, which reflected the will of the six signatory countries (Belgium, France, the Federal Republic of Germany, Italy, Luxemburg and the Netherlands) to set up a common market and establish an international structure: the European Economic Community (EEC). It was complemented by two other treaties: the Treaty of the European Coal and Steel Community, ECSC, in force since 1951, and the Euratom Treaty, also signed in 1957.

2 An avoided cost is the marginal cost or the cost a utility escapes by purchasing the product for resale from an IPP instead of building and operating a new plant.

Gradually, other countries joined the six founding ones, the total number of member States standing at 27 in 2013. Some key milestones of this construction were:

- the Merger Treaty of Brussels in 1965, that merged the executives of the ECSC, EEC and Euratom and established the European Commission and the European Council;
- the Single European Act, signed in 1986, which initiated the establishment of a common foreign policy in the 12 Member States;
- the Treaty of Maastricht, which established the European Union in 1992 and gave a new impetus to European integration;
- the Treaty of Lisbon in 2007, signed by 27 Member States in 2009, which helped change the institutional framework to adapt the EU to the enlargement which has taken place since the beginning of the 21st century.

The map in Figure 7.37 depicts Europe, with the names of countries in the local languages. With the exception of Albania, Bosnia Herzegovina, Croatia, Iceland, Kosovo, Norway, Serbia, Switzerland, Turkey and some principalities, the European Union includes in 2013 all European countries East of the Atlantic and North of the Mediterranean, and is bordered on the East by Belarus, Macedonia, Moldova, Russia and Ukraine.

Figure 7.37 Map of Europe. [*Source*: European Union].

Paradoxically, while two of the three original European treaties were related to energy (Euratom and ECSC), one has to acknowledge that for a long time there was no real European energy policy.

This probably stems from the contrast between the various European countries' energy situations, with some countries such as France, Belgium and Italy being much more dependent on imports than others, who have coal (Germany, United Kingdom), natural gas (Netherlands), and more recently oil (Denmark, United Kingdom) reserves.

In addition, until 1973 people lived in a period of apparent energy abundance during which joint European interventions did not appear necessary. After the oil shocks, governments preferred to deal with OPEC on a bilateral basis rather than through the EEC, and it is only by measured steps that the Commission was entitled to take an interest in energy issues.

As always in energy policy and as explained in Chapter 5, the balance is hard to find between the concerns of sovereignty and security on the one hand, and those of cost and efficiency on the other.

The actions of the European Commission in the energy field were initially mostly limited (apart from the actions under the two treaties mentioned above) to supporting the development of research on renewable energies, the rational use of energy and new energy technologies, actions that are still ongoing today.

A first revival of the work of the Commission followed the signing of the Single European Act, which allowed for the opening of borders with the establishment of a single market and a progressive liberalization of the electricity and gas markets from respectively 1996 and 1998 onwards. Given the differences that existed between the energy systems of individual countries, the Commission proposed to start to harmonize national situations in order to achieve a Europe of energy, mainly based on a liberal approach.

The basic idea calling for a liberalization of the energy sector was that the lack of a "Europe of energy" has a significant cost, which was assessed to amount to approximately 0.5% of the gross Community product for electricity only, the consumer having to pay this price. To reduce this cost, a shift to a wider competition appeared desirable, in turn making the global energy system more efficient.

Many obstacles however stood in the way of such a change, given the impact of national regulations and the existence of state monopolies. In a first step, a major transparency effort was planned, both in terms of pricing mechanisms and tax policy.

The Commission advocated for a greater transparency of gas and electricity tariffs and for the free movement of energy products, including the integration of gas and electricity networks. The idea was that through the payment of a toll to network owners, users could purchase energy from the producer or the supplier of their choice ("common carrier").

The situation has changed since the 1980s and Member States increasingly provide active support to the Commission. In 1986, the Council of Ministers of Energy has set some major goals for energy policy at the European level, and three important Directives have been adopted on public markets, the transit of electricity and energy price transparency.

It is, however, only since the adoption of the Treaty of Lisbon in December 2007 that the foundations of a genuine common energy policy for the twenty-seven Member

States of the European Union have been laid. This treaty gave this policy a sound legal basis (Article 194 of the Treaty on the Functioning of the European Union or TFEU), which had been lacking in previous treaties.

Adjustments to be made in Member States are still significant and will have a high cost, but it is only thus that a real European energy market may be created, which will ultimately benefit all economy sectors.

7.12.2 Outline of the European energy policy

The European Atomic Energy Community (Euratom) laid the foundations of Community cooperation in the civil nuclear field in 1957. Nuclear power plants currently generate approximately one third of the electricity and 15% of the energy consumed in the EU, but with great disparities between Member States. If Euratom has not given birth to a European nuclear industry, it has nevertheless led to cooperation in the field of research (ITER funding, etc.) and the development of safety standards.

More generally, a common energy policy became necessary for the Commission as well as the Member States insofar as the EU certainly has great advantages in terms of energy, given its technological and industrial expertise, but also presents vulnerabilities given its external dependence vis-à-vis fossil fuels. The share of imported energy in the EU, which already is, as we have seen, more than 80% for oil and 60% for gas, is expected to continue to increase in the coming decades.

Even if they are taken at a national level, the decisions of a Member State and the bilateral agreements it may conclude with non-EU countries also have an impact on the energy infrastructure and energy supply in the EU as a whole, which justifies a coordinated policy. A concerted action within the EU is generally more effective than an isolated one by a single Member State, in particular vis-à-vis producers and countries through which energy transits, in an environment of increased competitive access to resources.

As we shall see in Chapter 8, there is also a risk that the desire to preserve the environment, which has recently appeared among the top priorities, may only be addressed by appealing to more stringent regulations than in the past, which goes against the trend that sought to lower costs for the consumer, but likewise calls for a common policy of the EU Member States. The Union thus committed to reducing by 20% its greenhouse gas emissions by 2020.

Three options that appear essential underpin this policy: energy efficiency, the development of energy infrastructures and renewable energy.

One of the main objectives of the European energy policy is to accelerate the energy transition of the Union by promoting energy efficiency and the rational use of energy (including by eliminating waste) and developing new low energy consumption industrial technologies.

Significant reduction potentials do indeed exist, especially in energy-intensive sectors such as construction, manufacturing, energy conversion and transport. At the end of 2006, the EU committed to save 20% of its annual consumption of primary energy by 2020, including through the establishment of minimum standards for energy efficiency.

The development of the energy infrastructure is a second strategic European energy policy, with the double objective of opening up gas and electricity markets to

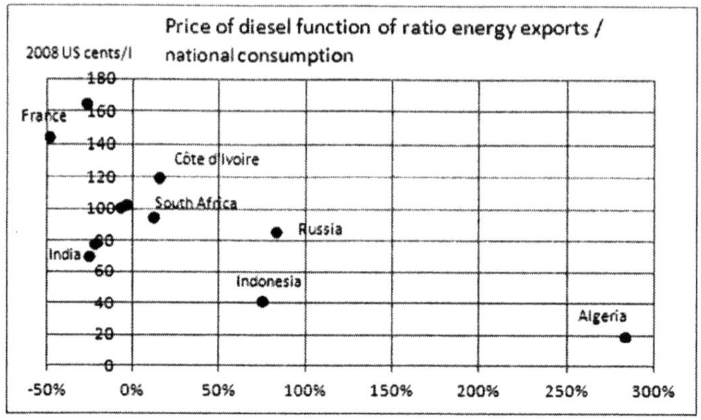

Figure 7.38 Diesel prices function of energy exports over domestic consumption.

competition so that the consumer may have the widest possible supply choices, and of strengthening the security of supply in the event of contingencies on supply routes, whether accidental, economic or political.

Finally, renewable energy, as we saw in Chapter 2, is an essential alternative to fossil fuels, which can additionally reduce greenhouse gas emissions from energy production and consumption. The EU therefore set itself the ambitious target of a 20% share of renewable energies in its 2020 energy balance. In the transport sector, which as of now depends almost entirely on oil, biofuels should account for 10% of total fuel consumption by 2020.

In December 2008 the EU adopted a package of measures aimed at reducing its contribution to global warming and securing its energy supply by strengthening international cooperation in this area, and has set a priority goal to achieve a genuine, better interconnected internal energy market, one that will therefore be less vulnerable to supply shocks.

In the previous paragraphs we mentioned the binding commitment made by the EU of reducing by 20% its greenhouse gas emissions and its primary energy consumption, as well as increasing by up to 20% the share of renewable energy in its energy mix by 2020. This threefold "20–20–20" objective is known as the European climate and energy package. By the end of 2012, the first and third objectives were well underway, while the second seemed more difficult to achieve, which led the Commission to consider corrective action plans.

Finally, note that the EU committed in 2009 to reduce by 2050 its emissions of greenhouse gas emissions by 80–95% below 1990 levels, which is a very ambitious goal. In the Energy Roadmap 2050 published in 2011 [European Commission, 2011], the Commission investigates via several scenarios the challenges set by the EU decarbonization goal. In addition to confirming the importance of the three options mentioned above, the Commission study concludes that the new objective can be achieved and will actually prove less expensive in the long term than the continuation of current policies: each scenario has benefits that can significantly reduce emissions, while being economically viable.

It is unfortunately not possible to further detail all ideas and actions that accompany the development of the European energy policy in this introductory book. The interested reader will find most of the information on the website of the European Commission given in reference.

7.13 CONCLUSION

We do not wish to synthesize these eleven countries' situations encountered by providing a systematic comparison of their indicators. Their principal characteristics are summarized in Table 7.1.

Let us only mention that the contrasts between some of them reflect the options they chose in terms of energy policy, including pricing. For example, Figure 7.37 provides the price of diesel as a function of the percentage of energy exports to domestic consumption, negative values being imports.

It illustrates that an abundance of resources often results in low tax or consumption subsidies, which can also lead to higher intensities.

REFERENCES

Analysis briefs of the United States Energy Information Administration: http://www.eia.gov/countries/

EC, Key figures, Market Observatory for Energy, Brussels, June 2011, http://ec.europa.eu/energy/observatory/countries/doc/key_figures.pdf

European Commission Energy Roadmap 2050, COM(2011) 885 final, Brussels, 2011 http://eur-lex.europa.eu/LexUriServ/LexUriServ.do?uri = CELEX:52011DC0885:FR:NOT

European Commission Energy Strategy for Europe, http://ec.europa.eu/energy/index_en.htm

National economic and energy statistics.

Vinois J.A. Ed., EU Energy Law – Volume VI – The Security of Energy Supply in the European Union, Claeys & Casteels Publishers, Deventer, 2012, ISBN: 9789081690423.

Chapter 8

Recent developments: New dimensions of the energy problem

8.1 PRESENTATION

The last few decades have been marked by numerous changes in the energy sector. Many of them have been mentioned in the previous chapters.

In this chapter, we will focus more particularly on two recent developments: the impact of energy on the environment and technological changes.

Lately, energy has indeed been directly involved in major issues concerning the environment, such as acid rains, the ozone layer depletion, the accumulation of CO_2 and the greenhouse effect, as well as nuclear accidents (Chernobyl, Fukushima) or oil spills. The threat of a generalized crisis of energy emissions that could jeopardize the current consumption patterns is drawing near. As we mentioned in the general introduction, this book does not aim at addressing all environmental issues. In the following pages, we will briefly analyze how this complex issue has evolved in recent years.

The second part of this chapter deals with the recent technological changes in the energy sector: while one could think that in this sector technological developments would likely be limited and slow, dramatic changes have occurred in recent decades, and these developments are still far from being over.

In the third part, we will look at developments in the supply and demand for oil and in particular the impact that technological progress has had on them.

8.2 ENVIRONMENTAL ISSUES

Since the early 1970s, environmental problems have begun to concern the general public. Neglected for a while, especially at the height of the oil shocks, these issues have since reappeared on the agenda.

The relationship of energy with the environment is complex and takes many forms, from the extraction of primary energy to its final use. Its aspects can be local, international or even global. Pollution is a major problem in industrialized countries, but also in developing countries, where the extensive use of firewood can additionally lead to the uncontrollable acceleration of deforestation, and thus desertification.

Generally, three types of pollution stemming from energy sources can be distinguished:

- soil pollution, due to the accumulation of coal, oil or nuclear waste;
- water pollution, from a physicochemical origin: coal mining and refining for freshwater, oil for the sea (offshore production, oil spills, ballast, etc.), from nuclear origin (accidental contamination of plant cooling water, reprocessing, etc.), or from thermal origin (plant cooling);
- air pollution, which is the main subject of current studies. It can take many forms: acidification, linked to emissions of SO_2 and NOx, climate change, linked to an increase in the greenhouse effect, radioactivity due to nuclear accidents, etc.

Some of these pollutions are accidental, others systematic. The only way to fight accidental pollution is to increase the systems' security and to develop depollution means so as to prevent human or technical failures that may have a disastrous impact on the environment. Concerning systematic pollution, the solution lies in the adoption of cleaner technologies, thanks to the addition of decontamination processes.

The decision of the EU countries in the early 1990s to require the installation of catalytic converters on automobiles to reduce emissions of SO_2 and NOx is an example: for catalysts to be effective, the fuel should not contain lead, which led to technical changes in engines (valve seats) and refineries. The adoption of such a measure thus has an impact on the entire energy chain.

8.2.1 Impact of energy on the atmosphere

The recent deterioration of a number of indicators of air pollution has generated global awareness of the need to closely look at developments in this area and seek solutions that are specially adapted to its transboundary nature, since existing regulatory frameworks generally do not provide relevant solutions.

8.2.1.1 Increased greenhouse warming

Similar to the mechanism used in a greenhouse or in a thermal solar collector, greenhouse warming is the influence of the atmosphere on the radiative balance of the Earth through the absorption of infrared radiation emitted from earth to space.

The greenhouse effect is caused by gases in the atmosphere, which absorb a portion of the incident radiation. The gases have absorption rays corresponding to different wavelengths. The diagram in Figure 8.1 gives a schematic indication of the effect of the absorption by CO_2 of the radiative flux of long wavelength emitted by the Earth, and shows that it absorbs two rays, centered on wavelengths $\lambda_1 = 4.5$ µm and $\lambda_2 = 14.5$ µm. The incident solar radiation of short wavelength (visible from 0.4 to 0.8 µm) is almost fully transmitted by the atmosphere and absorbed at the surface of the Earth, which heats and re-emits into space at 4 K radiations of long wavelength (infrared from 0.8 to 100 µm), an ever more significant part of which is absorbed by greenhouse gases. Their concentration having increased in recent decades, a gradual warming of the planet has been observed and could lead to significant climate changes.

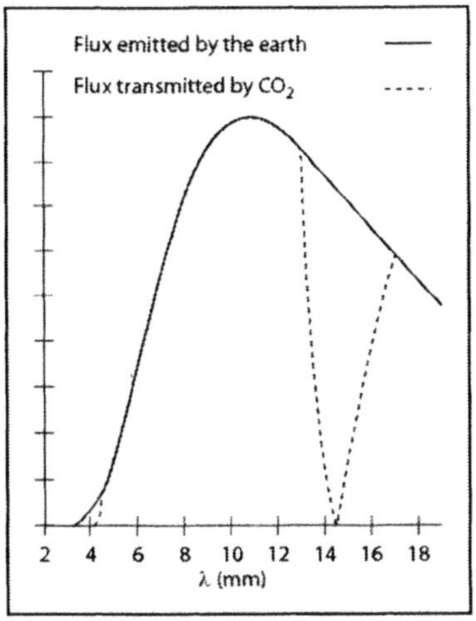

Figure 8.1 Radiation flux transmitted by the atmosphere.

Greenhouse gases are emitted as a result of many human activities (transport, heating, cooling, industry, livestock breeding, waste etc.), including the burning of fossil fuels (coal, oil and gas) that produces CO_2 that remains in the atmosphere more than a century on average before being recycled into "carbon sinks" such as vegetation. The main absorbing gases in the infrared range are water vapor H_2O, carbon dioxide CO_2, methane CH_4, nitrous oxide N_2O, chlorofluorocarbons (CFCs R11 and R12 in particular) and some noble gases. Their contributions to the increase of the (anthropogenic) greenhouse effect are highly variable (77% for CO_2, 14% for CH_4, 8% for N_2O and 1% for fluorinated gases), and depend on their concentration, a saturation phenomenon occurring beyond a certain threshold. This last point explains why the variations of concentration of water vapor have no significant impact on the global greenhouse effect, although water vapor is the most absorbing gas vis-à-vis the radiation.

Energy-related activities account for a significant portion of the greenhouse effect and its increase:

- the combustion of fossil fuels accounts for about 74% of CO_2 emissions due to human activities, the remainder stemming mainly from deforestation;
- the combustion of fossil fuels and biomass accounts for about 70% of anthropogenic emissions of N_2O;
- in the case of methane, the major share of the emissions comes from the fermentation of organic matter, the distribution and use of fuels being estimated to account for 10–30% of the total.

The distribution of greenhouse gas emissions in 2004 is given in Figure 8.2: 26% for the production of electricity and heat, 17% for land use activities, including deforestation, 14% for agriculture, 13% for transport, 8% for commercial and residential

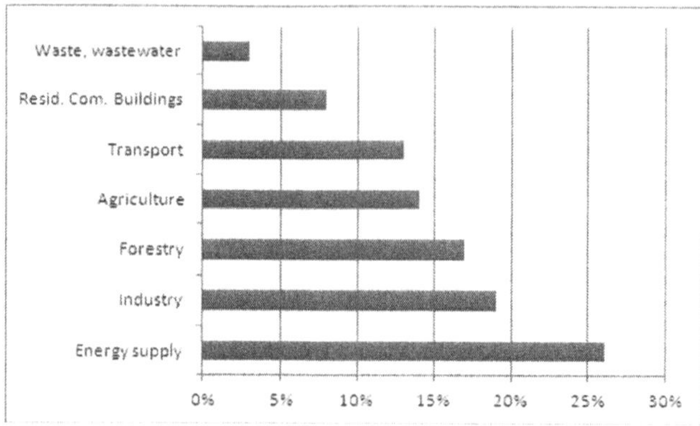

Figure 8.2 Sectoral distribution of greenhouse gas emissions.

buildings, and 3% for waste and wastewater. From the energy point of view, carbon dioxide emissions, given their preponderance, are now primarily being targeted.

Despite all the work that has been done over the past twenty years to better understand the mechanisms of the greenhouse effect, many uncertainties still remain, because of:

- the interactions between clouds, oceans, and polar ice caps;
- the size and nature of CO_2 sources and sinks.

In contrast, the share of the various gases emitted by human activities in global warming, sometimes called "radiative forcing", is better known. To evaluate the toxicity of a gas with respect to the greenhouse effect, an index known as the Global Warming Potential (GWP) is used. It represents the relative ability of a greenhouse gas to participate in radiative forcing, and is set at 1 for CO_2.

The contribution of a gas to the greenhouse effect depends on the one hand on the amount of emissions, and on the other hand on two factors taken into account in the calculation of GWP: its absorption properties in the infrared range and its residence time in the atmosphere. This last factor being very variable from one gas to another (several hundred years for CO_2, 10–15 years for CH_4), it is necessary to agree on an appropriate Integration Time Horizon (ITH). A gas's GWP can then be determined by calculating the cumulative radiative forcing on the chosen IHT. The IPCC (see below) has conducted assessments of GWP on IHT of 20, 100 and 500 years.

To assess the impact on the greenhouse effect of complex technologies such as those used for refrigeration, GWP is not sufficient. Indeed, refrigerants are usually not directly released into the atmosphere, while an indirect effect is due to CO_2 emissions corresponding to the energy consumed by the plant during its lifetime, which itself depends on the structure of the national electricity generation fleet. Another index was therefore defined: the Total Equivalent Warming Impact (TEWI).

For a refrigeration machine, TEWI takes into account the direct impact caused by the refrigerant charge m, the gas leakage during the lifetime n (annual percentage f)

and at the machine end of life (recycling factor α), and the indirect impact due to the energy consumption E – see equation (8.1).

$$TEWI = GWP \cdot f \cdot m \cdot n + GWP \cdot m\,(1-\alpha) + n \cdot E \cdot \beta \qquad (8.1)$$

$f \approx 5$ to 10%, $\alpha \approx 0.5$, $n \approx 15$, $\beta \approx 0.1$ kg/kWh for France, and $\beta \approx 0.55$ kg/kWh for Europe, the difference reflecting the importance of nuclear power in the French energy mix.

The calculation of equation (8.1) may seem a bit complex, but it makes it possible to compare very different technologies and to account for both direct and indirect effects. As shown in this formula, the value of TEWI can be decreased by reducing the refrigerant charge of facilities (m) and increasing their containment (f), which justifies the efforts being made in this direction by manufacturers.

Figure 8.3 shows the evolution of greenhouse gas concentrations from 1500 to 2010, that of CO_2 being expressed in parts per million and those of N_2O and CH_4 in parts per billion [MEDDE, 2013, after IPCC, 2007]. Note that the scale of the latter is on the right and that of the others is on the left. The increasing trend since the industrial revolution is very clear.

All models predict that emissions of greenhouse gases will continue to rise strongly in the near future, as they did in recent decades, and that their atmospheric concentrations will increase. The concentration of CO_2 in the atmosphere has indeed increased from 280 ppm to 360 ppm in a century, while it had remained within a range of 170 ppm to 280 ppm during the previous 200,000 years. Currently around 390 ppm, it is increasing by 0.4% per year and could exceed 500 ppm by 2050. Note that for the same energy consumption, CO_2 emissions vary in a ratio close to 1 for natural gas, 1.5 to 2 for heavy fuel oil, and 2–3 for coal.

Although, at present, carbon dioxide is primarily responsible for the increase in the greenhouse effect, the rate of emissions of other gases and their ability to retain infrared radiation are such that they could very quickly replace CO_2 in this role.

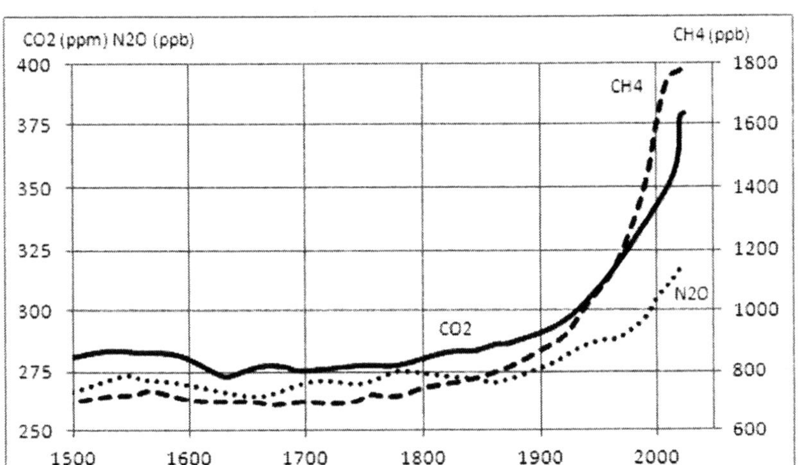

Figure 8.3 Greenhouse gas concentrations.

Scientific models are still unable to determine the real impact of these changes, one of the most difficult problems being to understand the behavior of the oceans, which are huge reservoirs of CO_2. One thing is certain, though: variations in global temperature and the concentration of CO_2 in the atmosphere are highly correlated.

If it therefore seems premature to draw definitive conclusions on the issue of CO_2, it is certain that great care is needed, and that alternatives should be considered now. Because CO_2 is an inevitable product of the combustion of fossil fuels, the only way to emit less of it is to burn less fuel. When you consider that fossil fuels represent nearly 90% of the global consumption of commercial primary energy, one can imagine the difficulty of the problem.

At the international level, the responsibilities are far from equal: three countries, the USA, Russia and China, are responsible for more than half the amount of CO_2 emitted worldwide, and 15 over 80%. Emissions per capita and per year are highly variable (18 tons of carbon in the USA in 2011, 12 in Russia, 8 in the UK, 6 in France, 5 in China, 2 in Brazil, 1 in India, and 0.35 in Côte d'Ivoire, the world average being equal to 4.4).

Global warming has now become a major environmental issue, and is the subject of national and international policy negotiations.

The international community responded in 1988 by creating the Intergovernmental Panel on Climate Change (IPCC), under the United Nations Environment Program (UNEP) and the World Meteorological Organization (WMO). The IPCC is charged with evaluating the scientific evidence on climate change, assessing its ecological and socio-economic impacts, and proposing strategies for prevention and adaptation.

IPCC reports confirm the influence of greenhouse gas emissions on the Earth's climate, already demonstrated by many recent statistical anomalies, like the El Niño phenomenon, which causes unusual droughts and floods in Latin America. Scientists have found that greenhouse gases have been responsible for a 0.6 to 0.9 °C increase of average temperature at the Earth's surface since the preindustrial era, which has been partially obscured by the thermal inertia of the oceans and the increasing presence of aerosols in the atmosphere. It is estimated that the average temperature at the Earth's surface could rise by another 1.8 to 4 °C by 2100, resulting in a 18 to 60 cm rise of sea level.

Although the local climatic consequences of the increase in the greenhouse effect are not well known, IPCC experts believe that major disruptions are inevitable, with an increased frequency and intensity of natural disasters (droughts, floods, storms, cyclones).

Although not absolutely proven, many scientists believe that global warming contributes directly to increasing the destructive potential of cyclones, as evidenced by Katrina, which devastated Louisiana in 2005, or Sandy that hit the East Coast of the United States in November 2012, leading to the evacuation of 375,000 people in New York, plunging half the city into darkness and killing 210 people.

The United Nations Framework Convention on Climate Change (UNFCCC) was ratified in 1992, following the Rio de Janeiro Summit. Under this agreement, effective March 21, 1994, the OECD countries, Ukraine and Russia pledged to stabilize their emissions of greenhouse gases by the year 2000 at the same level as in 1990. In December 1997, at the Kyoto Summit, which led to the Kyoto Protocol entering into force in February 2005, these commitments were extended for the period 2008–2012:

the industrialized countries pledged to reduce by an average of 5.2% their emissions of greenhouse gases over the 2008–2012 period as compared to 1990, the European Union for its part undertaking to reduce its emissions by 8%. As part of an agreement within the European Union, France must stabilize its emissions (0% target).

With the first phase of the Kyoto Protocol expiring in 2012, international negotiations resumed at COPs (Conferences of the Parties) in the late 2000s in order to establish targets to reduce emissions up to 2020. The last COP at the time of writing this book, that of Doha in 2012, only resulted in a partial extension of the Kyoto Protocol until 2020, several "Annex 1" countries (Canada, the United States, Japan, New Zealand, as well as Belarus, Kazakhstan, Russia and Ukraine) having announced that they would not participate in this second phase.

In a context of steady economic growth, in which 85% of energy needs are provided by fossil fuels, forecasts are predicting a doubling of current CO_2 emissions by 2050 if no changes are effected, i.e. 50 billion tonnes emitted per year. Stabilizing the CO_2 concentration in the atmosphere around 550 ppm would require cutting by half the emissions compared to 1990. In the light of current technological and economic constraints, this goal is a truly global challenge of unprecedented scale.

Indeed, even if the need to reduce emissions of greenhouse gases seems to be recognized by the implementation of national and international public policies, the major difficulty is the strong link between CO_2 emissions and economic growth. All developed economies consume a lot of energy, and especially fossil fuels.

Two sets of indicators of policy performance can be used, depending on whether one is interested in per capita values or intensities relative to GDP:

- per capita energy consumption;
- per capita carbon emissions $ (CO_2 emissions/capita);
- the energy intensity of the economy, as defined in Section 3.6 (energy/GDP);
- the carbon intensity of the economy (CO_2 emissions/GDP).

These indicators allow one to characterize the types of mitigation measures, a set of measures being needed to build an effective and economically acceptable policy.

8.2.1.2 *Depletion of the ozone layer*

Since English and Japanese researchers demonstrated the existence of a hole in the ozone layer over Antarctica during the austral spring (a decrease by half to two thirds of the normal thickness) in 1985, the international community has been particularly active on this front.

The depletion of the ozone layer is due to many gases, particularly those containing chlorine, as this element reacts with ozone according to the reactions:

$$Cl + O_3 \rightarrow Cl\,O + O_2$$
$$Cl\,O + O_3 \rightarrow Cl + 2\,O_2$$

Such gases include chlorofluorocarbons (CFCs), whose use was at that time about 60% directly related to energy applications: production of thermal insulating foam and use as refrigerants in air conditioning and refrigeration.

The first international agreement concluded in this respect was the Montreal Protocol, which took effect in early 1989, and decided to freeze CFC consumption at 1986 levels, followed by a reduction of 20% in 1993, and a second reduction of 30% in 1999. The move then continued, and in 1990, an agreement was reached in London on an amendment resulting in an acceleration of the process, with the goal of outright banning of all CFCs and halons from 2000 on, and a calling into question of HCFCs, since they still contain chlorine.

The European Commission Regulation of March 1991 demanded that production and consumption of CFCs be stopped by 1 June 1997. In November 1992, in Copenhagen, a new amendment brought this date forward to 1 January 1996.

The way in which the ozone layer depletion issue was tackled at the international level is particularly exemplary: after some hesitations, far from adopting a conservative attitude, manufacturers generally supported mandatory regulations.

This is not common and the reasons for this behavior are numerous. However, it is clear that this threw manufacturers into a technological race where those who had invested in research substitutes for CFCs sought to capitalize on their advance by advocating a strict regulatory framework, thus limiting the market share of their competitors. Behind the apparent convergence of views there was a battle between a few large industrial companies in industrialized countries and their counterparts in the "South", particularly in India and China.

It is impossible to review all options and constraints with respect to the issue of replacement refrigerants in this book. We will therefore only make a fairly general presentation, referring the reader interested in further developments to the specialized literature.

Reducing the ozone depletion potential of the used fluids implied a modification of their thermal and thermodynamic properties, which led to the questioning of many previous technological solutions, and posed many challenges to refrigeration and air conditioning industries.

The issue is different if a new installation should be designed or if only the fluid should be changed in an existing one. In the latter case, it is necessary that the thermodynamic properties of the replacement fluid are similar to those of the original, whereas in the first case the change can sometimes lead to improved performance of the facility. In any case, many technological problems must be studied, such as the compatibility with lubricants, the choice of the dehydrator, the adjustment of the expansion valve etc.

To complicate things, regulations differ between countries, depending on the commitments they took on under international agreements. A final element to consider is that great uncertainties remain for the future, especially as regards the fate of high-GWP HFCs.

The evaluation of the toxicity of a gas in terms of destruction of the ozone layer can be estimated using an index called the Ozone Depletion Potential (ODP), whose value is equal to 1 for gas R11 (CFC).

CFC replacement fluids can be grouped into three broad categories: transition fluids, zero ODP fluids, and zero ODP fluids with low GWP.

The first solution that came to mind to replace CFCs were hydrochlorofluorocarbons (HCFCs), *a priori* relatively close on the chemical level and less harmful to the ozone layer. However, these non-zero ODP fluids are either already banned or will soon be by the signatory countries of the Montreal Protocol and the agreements that followed, so that they are only a short-term solution, hence their name: "transition fluids".

In the longer term, the only halogenated fluids that are acceptable vis-à-vis the ozone layer are hydrofluorocarbons (HFCs), which do not contain chlorine atoms and have a zero ODP value.

The main drawback presented by these fluids is that while their ODP is zero, their GWP is very high (1,300 for R134a, 3,200 for R125, 580 to R32, 4,400 for R143a), and their molecules, very stable, have atmospheric lifetimes of several decades or even hundreds of years. Their contribution to the greenhouse effect is thus potentially significant, and their production is therefore likely to be questioned sooner or later.

The only alternative fluids that do not present drawbacks regarding either the ozone layer or the greenhouse effect are non-halogenated fluids such as ammonia (R717), propane (R290), isobutane (R600a), carbon dioxide (R744) and water (R718).

The first three have flammability constraints (as well as toxicity and compatibility issues with some metals such as copper does for ammonia) that limit their use. The latter cannot be used for negative temperatures, whereas R744 cycles have efficiencies significantly lower than the others.

Ammonia, butane and propane having very good thermodynamic characteristics, they represent serious alternatives, and their use has grown significantly in recent years. 30% of new European domestic refrigerators now use butane, particularly in Northern Europe.

Despite its relatively low performance and the high pressures that it requires, carbon dioxide is under consideration as a fluid for vehicle air conditioning systems, where the high risk of leakage challenges the use of high-GWP R134a.

It is also possible to use blends as replacement fluids. In fact, manufacturers of refrigerants consider that with the exception of R134a, it is now unlikely that pure fluids will be discovered whose thermodynamic properties would allow them to perfectly replace the old ones. However, by mixing pure fluids in well-chosen proportions, it is possible to obtain more suitable features, which explains the interest in the blends.

8.2.1.3 *Acid rains*

Acid rain is a phenomenon that leads to the degeneration of forests, especially in Central and Northern Europe. The share attributable to energy is far from being clearly established, but it seems obvious that it plays an important role, because of emissions of sulfur and nitrogen oxides that are released during the combustion of coal or oil.

The combustion of coal and oil emits sulfur dioxide (SO_2) and nitrogen oxides (NOx), which, dissolved in the water in the clouds, make it acid. It is estimated that, when it comes to energy, sulfur dioxide is responsible for 65% of the emitted acidity and nitrogen oxides for about 30%.

The energy uses responsible for acid rains are firstly electricity generation and transport, and to a lesser extent the heating of facilities. The fight against acid rains in Europe has thus resulted in the decision to equip motor vehicles with catalytic converters, and to strengthen regulatory measures for the production of electricity from coal.

In large cities, public opinion has been alerted by the fact that concentration thresholds have been exceeded for some pollutants during peak pollution. Governments are therefore finding it increasingly necessary to reduce pollutant emissions in vehicles and energy production facilities.

Regarding transport, electronic control injection systems helped make significant advances in reducing pollutants (50-fold decrease in 30 years and 10-fold over the last twenty years) but they are insufficient to meet the new requirements imposed by international regulations.

After consultation with the manufacturers, the European Union issued guidelines setting a number of goals. In particular, Directive 2001/81/EC adopted on 23 October 2001 demanded that Member States establish a national program to reduce emissions of SO_2, NOx, volatile organic compounds (VOCs) and NH_3, to meet by 2010 the limits set for each of these four pollutants. The aim of the Directive was for Europe to meet its commitments under the Convention on Long Range Transboundary Air Pollution, adopted in Geneva in 1979, and the Protocol on the reduction of acidification, eutrophication and ground-level ozone, signed in Gothenburg in 1999.

Measures to reduce emissions of SO_2 and NOx comprise the widespread use of catalytic converters for automobiles, and, for thermal power, prior desulphurization, as well as new combustion techniques (fluidized bed, the addition of limestone, etc.), desulfurization and dust removal.

The dotted curve in Figure 8.4 shows, between 1990 (index 100) and 2005, the reduction in sulfur dioxide emissions in the countries of the European Environment Agency, which included 32 members in 2012. These emissions decreased by 66% during this period, the two sectors with the highest emissions being the energy sector (41%) and industrial processes (35%).

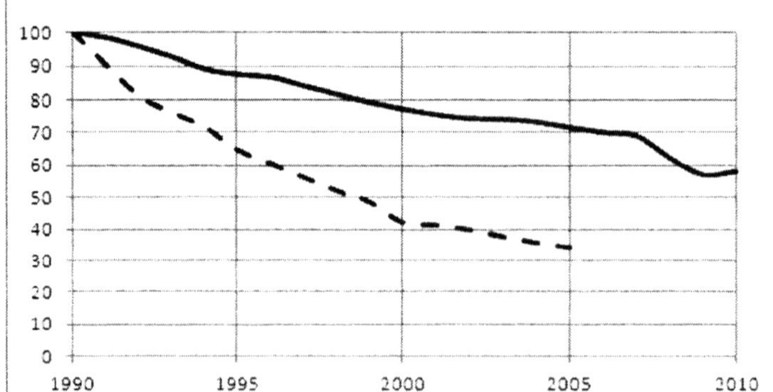

Figure 8.4 Decrease of SO_2 and NOx emissions in EU-32.

However, emissions of nitrogen oxides (solid curve until 2010) could not be reduced as much (only 28%), with transport counteracting gains in other sectors. The share of transport in NOx emissions represented 40.5% in 2010, and the energy sector 22.4%.

8.2.2 General remarks on additional environmental impacts of energy

On the whole, a study of the environmental impact of the various kinds of energy leads to the following conclusions:

Solar thermal or photovoltaic energy poses no particular environmental problem besides the land area required for the facilities and the aesthetic qualities of collectors.

Geothermal energy, except for open deep cycles, also has a negligible impact on the environment.

Biomass should have a modest impact, with the reservation made above for deforestation, and the consequences in terms of CO_2 emissions.

Hydropower has a direct impact on the flow regime of water in dam sites: alluvial deposits, groundwater, etc., with induced impact on flora and fauna. In addition, there is a risk of accidental breakage of the dam, which can be reduced by geological, seismic and mechanical studies.

Nuclear power generates nuisances of variable importance. Thermal pollution of water is now well understood. Radioactive emissions in normal operation are a well-controlled problem, although there is still room for progress regarding the doses received by emergency personnel.

The nuclear waste problem is practically solved for products with a low activity and a short life, while geological solutions should arise in the next century for products with a high activity and a long life.

The most critical problem is the safety of nuclear reactors themselves. The Chernobyl accident in 1986 was directly responsible for 31 deaths, the evacuation of 135,000 people in the fallout zone extending 30 km downwind of the plant, radioactive fallout over much of Europe, and could be the cause of tens of thousands of instances of cancer in the next 50 years.

The Fukushima disaster on 11 March 2011 saw the combined effects of a nuclear accident and of an earthquake (magnitude 9) followed by a tsunami. It is too early to know precisely the casualties it caused. An estimate of the number of deaths directly caused by the nuclear accident is 573, and the emissions are considered much lower than those of Chernobyl (10 times less iodine and 3 times less cesium).

Fossil fuels, coal, oil, and to a lesser extent natural gas, are seen as the main cause of the increase of CO_2 in the atmosphere and acid rain. To this environmental pollution common to various fossil fuels must be added the specific problems of each source: mine accidents, silicosis, dumps, oil spills, etc.

Increasingly stringent oil exploration and production conditions result in more severe threats on the marine environment, as shown by the April 2010 Macondo drilling accident in the Gulf of Mexico: the sinking of the Deep Water Horizon oil rig due to pressurization of the wellbore resulted in the death of eleven people and unprecedented marine pollution.

8.2.3 Possible solutions

To minimize the negative impacts of the energy sector on the environment, three types of solutions can be envisaged:

- the rational use of energy and improvement of energy efficiency (energy conservation), which should be considered a priority at all levels of the energy chain. Reducing the consumption or slowing its growth will improve both the availability of energy and the environment;
- the use of clean technologies, whenever possible, through the choice of cleaner processes and energy sources emitting less carbon than oil (natural gas, nuclear energy, wind, etc.), the installation of processes that reduce emissions during combustion, primarily for SO_2 and NOx, and the use of depolluted products;
- finally, the implementation of means to prevent and repair damages has to be systematized: rehabilitation of mining sites, sewage, waste disposal, CO_2 capture and storage, etc.

According to the "polluter/payer" principle, it is now increasingly recognized that the costs of pollution prevention must be borne by the person who pollutes. This principle, endorsed by the OECD in 1972, is the basis of the environmental law of many countries and in particular that of the European Union.

When it is possible to establish beyond any doubt the exact responsibilities of polluters, they may be charged abatement costs, especially in the form of (Pigouvian) taxes. However, this is rarely the case and there is no method to optimize decisions at the global scale. It is moreover also difficult to assess the cost of the generated pollution.

The introduction of an international carbon tax therefore seems unlikely, and markets for emission allowances have been developed instead. The Kyoto Protocol, mentioned above, allows "Annex 1" countries, which are committed to reducing their CO_2 emissions, to use three "flexibility mechanisms" to achieve their targets:

- the emissions trading mechanism: a state or a company that emits more than its quota can buy emission allowances (carbon credits) from an entity whose emissions are below its allocated quota; a European market, the EU ETS (European Union Emissions Trading Scheme) has been established in 2005;
- the Clean Development Mechanism (CDM), which allows industrialized countries to receive emission credits in return for financing low-carbon projects in non-Annex 1 countries;
- the joint implementation, similar to CDM but for projects realized in other Annex 1 countries.

Precise control (measuring, reporting, verification or MRV) is necessary for these mechanisms to be effective. Emission markets are indeed subject to criticism: in addition to numerous suspected frauds, some observers regret their limited incentives for emission reductions for the most polluting industries, especially in a context where the price of carbon credits has collapsed in recent years.

In any case, it seems that the only effective way to compel manufacturers to be more environmentally friendly is through regulation and supervision. Given the international nature of the most serious pollutions, the regulation must be developed at this level. Note that the need to strengthen pollution standards may conflict with the trend towards more liberalism advocated by the promoters of a European energy system.

Generally, in terms of impact on the global environment, it should be noted that the knowledge of the involved mechanisms and their extent is still in its infancy, despite progress made in recent years. It is therefore expected that regulations will change significantly in the coming years, as constraints and degrees of freedom appear more clearly.

8.3 CHANGING TECHNOLOGY

8.3.1 Introduction

While one might have thought that the energy sector would have experienced only limited and slow technological developments, dramatic changes have occurred in recent decades, and this process is far from over. In this section we will try to highlight the reasons behind this apparent paradox by considering briefly the overall social context where the technology is used, which is marked by a number of characteristics:

- energy being ubiquitous in modern societies, the financial stakes are considerable. The competition is very open as many technological solutions exist to satisfy a given final energy need.
- for engines many technology solutions are in strong competition, used in particular for propulsion and power generation. The choice depends on several criteria, including of course the specific type of use, the available energy sources and their cost, emission regulations etc.;
- for refrigeration cycles, two technologies compete: mechanical vapor compression and liquid absorption. Because the latter remains quite marginal, it will be only briefly reviewed here. The main problem encountered for vapor compression cycles is that the vast majority of them used refrigerants that destroy the ozone layer until recently but are now banned by international agreements, as we have seen above;

The energy sector is generally considered technologically mature and not prone to experience severe and rapid technological developments, on the one hand because of the rigidity of its sub-sectors and on the other the long use life of the equipment used. One might therefore conclude that technological innovation can only play a secondary role in the global energy problem. The analysis of changes over last 30 years shows that this is not true and on the contrary that technological innovation could be the key to our energy future. During this period, many limits deemed impassable were exceeded, and many records established.

Advances in oil exploration and production have led to the exploitation of new deposits in countries outside OPEC. The new oil supplies and reductions in consumption due to various substitutions and savings have played a decisive role in upsetting the oil market as was experienced in 1986. Horizontal drilling techniques now make the exploitation of unconventional oils possible. The development of combined cycles allows countries to produce electricity with an efficiency approaching 60% today, while 40% seemed economically unattainable before.

Through energy conservation, energy intensity, which represents the amount of energy needed to produce one unit of GDP, fell by 30% in 20 years in many OECD countries.

Meanwhile, the unit quantities of pollutants from energy processing and equipment use fell sharply, and this trend should continue in the future. Even when international energy prices were relatively low, the development of new energies such as controlled fusion, fuel cells, solar technologies etc. continued around the world.

If we look at what the main drivers of technological change in recent decades were, three points should be specifically discussed.

The first driver (at least chronologically) was the fact that fossil fuel resources are limited and their distribution uneven in the world, resulting in a random availability and thus justifying the search for alternatives.

The second driver of technological change – in chronological order, as it is today the paramount driver – is the environmental impact of energy technologies, as we mentioned at the beginning of this chapter. Indeed, even if non-renewable energy were available without restriction, the total volume of energy technology pollutant emissions has become so significant that current consumption patterns would be questioned.

The third driver of technological change is the progress made in other fields that have a direct impact on the energy sector.

Three of them are worth mentioning here, the first two being related to information and communication technologies:

- theoretical and numerical modeling and computerized codes have played and will continue to play a fundamental role in technological change. Progress in software and increased computing power make once unthinkable modeling and optimizations possible that improve the performance of many energy technologies;
- in a related field, progress in real time electronics and control allows operators to pilot facilities according to protocols that are much more accurate than before and to improve their reliability;
- progress on materials also has a direct impact on energy technologies, e.g. the protection of gas turbine blades, catalysts for exhaust converters, etc.

8.3.2 Exploration and production in the oil sector

After the first oil shock, oil companies invested heavily (increased budgets for R & D by factors of up to 4) in programs to diversify oil supplies in a timely manner, which enabled them to discover new oil fields outside OPEC and to develop quite remarkable new technologies, as for example in offshore fields.

In 1986, as prices had fallen, oil companies had to face another challenge: the additional costs were no longer acceptable. However, as the effectiveness of technological innovation had been demonstrated, the R & D programs were generally not questioned: they were redirected towards a new goal, i.e. the reduction of costs to maintain and renew oil production outside OPEC. Of course, the recent surge in oil prices puts all the technological advances in the oil and gas sectors back on the agenda.

In exploration-production, this resulted in the development of the most promising technologies among those which had previously been designed, such as 3D reflection seismology and drilling tools and techniques.

3D reflection seismology allows:

- firstly, in exploration, thanks to a much better representation of the ground, new structures to be revealed and to drill with success rates well above the previous standards;
- secondly, in development, a better appreciation of the flow's physical properties and of the nature of the fluids contained in reservoirs;
- finally, in production, to better follow the evolution of the field and to guide the drilling of complex and large deflection wells, leading to a sharp increase in the recovery rate.

Drilling tools and techniques have evolved in recent years, with:

- reduced costs (slim hole, coiled tubing);
- deviated drilling, which allows to develop increasingly complex well architectures and to better drain reserves. Offsets greater than 10,000 m have already been achieved;
- multiphase pumping, for deep offshore production (especially beyond 1,000 m of water).

All these technological advances have yielded very significant results:

- a very strong reduction of technical costs in the oil industry, which is expected to continue in the future;
- a large increase in the rate of oil recovery, which can reach 60% in some fields, while it was limited to 40% 25 years ago. The global average recovery rate has increased from 25–30% in 1985 to 30–35% today;
- a reassessment of world oil reserves (between 1982 and 1992, additional oil reserves corresponded to 4.4 Gt for new fields, and 4 times more or 19 Gt for revisions and extensions of fields already known);
- the discovery and introduction of new oil basins in different regions (South America, Africa, Asia-Pacific, Western Europe, CIS).

8.3.3 Improved combustion

Aware of the growing challenges associated with reducing sulfur and nitrogen oxides, energy technology manufacturers are investing more in search of a better control of combustion conditions, particularly for engines and electricity power generation.

Let us simply indicate at this stage that the problems are very different in conventional thermal power plants, where the combustion chambers are large, the pressure being close to the atmospheric one, compared to internal combustion engines, where combustion takes place in much more compact volumes, with increasingly brief residence times and under high pressure.

8.3.3.1 Electricity production

By way of illustration, let us compare the size of two typical 250 MWe power plants, one using a steam cycle with conventional pulverized coal boiler, the other a modern gas turbine:

- the burner of a coal boiler typically has a single ground base of 11 m × 11 m and a height of 27 m, a volume of 3,200 m^3; its net efficiency is around 39%, which corresponds to a thermal power of 250/0.39 = 641 MW;
- a gas turbine (GT 26 ALSTOM) has a capacity of 260 MWe and an efficiency of 36% for a combustion chamber volume of only 4 m^3, which corresponds to a thermal power of 260/0.36 = 722 MW.

The thermal power released per m^3 of combustion chamber is about 200 kW/m^3 for the coal plant, and 180 MW/m^3 for the gas turbine, i.e. 900 times larger. This very important difference comes from the fact that in the case of boilers, there is a barrier (mechanical and thermal) between the fumes and the working fluid, which limits the heat transfer, whereas in the gas turbine, energy is released directly into the working fluid.

8.3.3.2 *Car engines*

Remarkable results were also obtained in terms of improving the efficiency of automobile engines, especially in Europe, with a particular emphasis on limiting their impact on the environment.

Direct injection is not a new concept, far from it. Most semi fast or slow diesels use it. In the fast engines, it was not used until recently because it had several drawbacks (particulate emissions due to the heterogeneity of fuel distribution, engine noise etc.) and did not accelerate enough combustion. The solution adopted was to use pre-chambers, which increased the losses through the walls and throttling.

The work on direct injection was initially focused on diesel, the aim being to avoid the problems that previously existed when used in fast engines. The obvious solution today is the Common Rail technology, which appears the best compromise on the technical-economic level. Roughly, the fuel is distributed through a pipe common to all the injectors, which are steered through electromagnetic devices. The very high "rail" pressure (1,300–1,500 bar) and the very small diameter holes of the injectors (less than 200 microns) makes it possible to achieve extremely fine fuel atomization, so that it immediately vaporizes and burns much better in the air than was possible with conventional systems.

Diesel engine performance could thus be much improved compared to those of indirect injection engines of the previous generation:

- 20% reduction in consumption (and thus CO_2 emissions);
- a decrease of 40% of CO, 50% of unburned hydrocarbons and 60% of particulates at a constant level of NOx;
- gains in drivability (over 50% increase in torque at low rpm), reduced noise and vibration.
- This progress is attributable partly to direct injection, and partly to the engine downsizing (12%) and the reduction of internal friction (reduced consumption of 1.5% to 2% at low and medium speed).
- Similar to developments made for diesel, direct injection is being developed for gasoline engines, with two technologies in competition: lean-burn and stratified charge, such as developed in Japan since 1997 by Mitsubishi and Toyota, and high stoichiometric Exhaust Gas Recirculation (EGR), which correspond to the

choice made by Renault, which has equipped the Mégane with the first models since 1999.

The goal here is not only to optimize the nominal performance of the engine, but also to reduce losses at low speed due to pumping and arising from the impossibility, in a conventional gasoline engine, to ensure complete stable combustion below the flammability threshold of the fuel mixture (air/fuel ratio close to 22/1, while the stoichiometry is achieved for 14.6/1). The idea is to circumvent this difficulty by achieving, thanks to the injector, a heterogeneous mixture in the chamber, but of suitable composition near the spark plug, so that combustion may then spread to the rest of the chamber.

Figures 8.5 and 8.6 illustrate fuel economies that are foreseen for both gasoline and diesel engines in the following years.

Thanks to progressive innovation, fuel consumption and CO_2 emissions of diesel and gasoline engines are expected to decrease by about 30 percent by 2015.

8.3.4 Production of electricity from natural gas

- Recent developments in gas turbines and combined cycles offer a striking example of results quickly achieved through technological innovation in the energy sector. These technologies, until recently still regarded as relatively marginal, are poised to play a leading role in the structure of power plant fleets. They are much less sensitive to scale effects than conventional power plants, for which the increase in size was a competitiveness factor. Factory-made, they come in modular form and are assembled on-site within a much shorter time. The benefits are numerous;
- it is possible to stagger investments, thereby reducing financial costs;
- the footprint is smaller than that of conventional power plants;

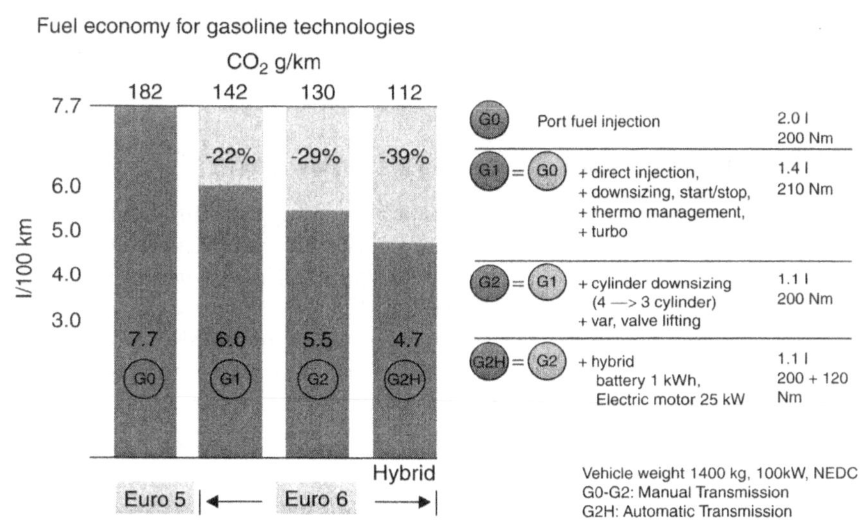

Figure 8.5 Comparison of concepts designed for gasoline engines, Source Bosch.

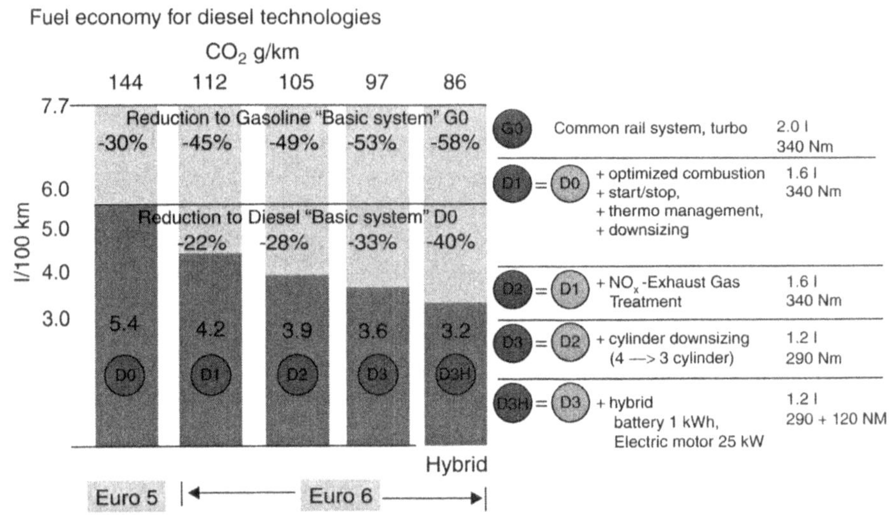

Figure 8.6 Comparison of concepts designed for diesel engines, Source Bosch.

- a large power unit is composed of several modules in parallel, thereby facilitating maintenance;
- the difference between centralized and decentralized production is reduced, which helps diversify the production sites and thus reduces network vulnerability.

As discussed in section 2.5.1, recent progress on ground machines is largely the result of a technology transfer from the aviation world. In fact, most jet engines used in aviation today are variations of the open cycle gas turbine. For 50 years, the tremendous growth experienced by the aviation market has helped fund major technology development programs on these engines, which led to the development of highly efficient and competitive gas turbines.

8.3.5 Reduction of specific energy consumption

It is important to note that we can reduce primary energy consumption while increasing the satisfaction of useful energy needs. It is because significant losses exist in two main steps of the energy chain: the transformation of primary energy into secondary energy and final use, as we have seen in section 3.5.

It is perfectly possible under these conditions to significantly increase the useful energy available while lowering the consumption of primary energy. To this end, efforts should focus on the one hand on optimization of the supply system (refineries, and especially power generation) and on the other hand on improving energy-using equipments.

Demand management schemes are aimed at achieving this second objective. The establishment since 1973 of resolute policies for energy conservation yielded significant results in terms of reducing energy intensity, including in industrialized countries, to the surprise of many observers. Energy efficiency has today become an essential component of any energy policy. Let us examine the main results.

We mentioned earlier that, before the first oil shock, forecasting energy demand in a country was commonly based on the assumption of a constant energy-income elasticity, around 1 in industrialized countries. If we recall the mathematical definition of elasticity, $e = 1$ means $c = p$ or $i = 0$, following the notations in Chapter 3.

In other words, during this period, energy intensity maintained a constant value, which suggests that the structure of GDP as well as the specific equipment consumption varied little. At the time, very few observers imagined it might be possible to decouple the growth of the economic activity from that of energy consumption by acting proactively on demand.

In fact, a closer examination reveals that the share of the manufacturing sector in the overall energy demand fell slowly due to the growth of the tertiary sector, while the energy consumption of households increased moderately as a result of rising standards of living (household equipment, comfort, transport) and decohabitation (family breakdown). For example, the equipment rate of French households increased from 1960 to 1988, from about 30% to 75% for cars, 25% to 97% for refrigerators, 0 to 40% for freezers, 10% to 94% for television, 30% to 86% for washing machine, and the trend has continued since, so that in 2007, almost all French households had a refrigerator, a TV and a washing machine. The proportion of households with at least one car increased slightly for several years, while the share of those who have at least two has doubled since 1980: over 30% of households have two cars and 5% of them have three or more cars, against respectively 15% and 2% in the early 1980s.

We showed in Sections 3.6.5 and 3.6.6 the changes in GDP, energy consumption and energy intensity in France and in OECD (Figures 3.8 and 3.9).

Reducing energy intensity is the result of two complementary changes:

- a change in the structure of the economic activity, because of the development of the low energy intensity tertiary sector (structure effect);
- a reduction of the specific consumption of energy processes and equipment, as shown below for some products.

In the industry, the specific energy consumption decreased in almost all branches: textiles, ammonia production, laminates, glass, asbestos, etc. Moreover, oil consumption fell by 60% from 1973 to 1988, and was replaced by coal in cement plants, paper mills, food processing, and heavy chemicals, and by natural gas heating for various uses or production of decentralized steam. It then continued to decline from 20.3% of the total in 1993 to 13.8% in 2009.

Efficient electrical techniques became widespread: mechanical vapor recompression, membrane filtration, plasma torches, induction heating, infrared heating, etc.

In the housing sector, the average energy consumption of new homes has been declining since 1974 as a result of different steps of thermal regulation: 1.4 toe/year for a housing constructed according to the standards of 1974, 0.85 toe/year for a post-1987 housing benefiting from the two stars High Energy Performance (HEP) label. More recently, the criteria have been further strengthened and diversified to promote the use of renewable energy. Thus the French thermal regulation "Grenelle Environnement 2012" has set as a target of primary consumption 50 kWhoe/m^2/year, or 0.43 toe/year for a 100 m^2 dwelling.

In the transport sector, thanks to the fabrication of more energy efficient vehicles, the average consumption of French gasoline cars decreased from 7.5 L/100 km in 1995 to 6 L/100 km in 2008 and that of diesel vehicles from 6.6 to 5.3 L/100 km, which is in both cases a gain of 20%.

Large energy savings have also been made on household appliances in Germany since 1975: more than 60% for washing machines and freezers and more than 40% for electric furnaces. Neon bulbs that provide 11 W illumination equal to that of a 60 W filament bulb were developed, which represents a gain by a factor of 5.

These changes are slow because it takes more than a hundred years to completely renew housing in a country and more than twenty years to completely renew the car fleet. The effects of the implementation of demand management policies are very progressive, but they are no less real.

These savings stem from technological advances, resulting in both the development of high-performance materials or techniques that did not exist a few years ago and an improved use of existing equipment.

In this dual effect (new products and better use of existing ones) lies the rich potential of energy conservation: it allows one to achieve an expected profitability for the corresponding investments, although sometimes financial assistance from the State may be needed as the consumer micro-level optimum may differ from the macro optimum.

For the consumer, payback time has been extremely short for these technological substitutions:

- a particularly significant example is the introduction of deflectors on trucks for transport, whose investment is amortized in less than six months;
- recovery of industrial waste heat represents a potential savings of several million toe, and this potential is only partially exploited today;
- in housing, the expected energy savings brought about by the generalization of efficient isolation techniques are also very substantial.

8.3.5.1 Considerable scope for future progress

Besides these abovementioned examples of energy savings, which are achievable in the short term, technological research programs currently underway will allow, in the longer term, for further lowering of the energy consumption in different sectors.

The vitality of the research on energy conservation foreshadows the emergence of new techniques that could be applied in the near future and that justify the efforts that are devoted to it. However, it is clear that there are limits to increasing the overall efficiency of the energy system. These are physical (derived from the second law of thermodynamics) as well as economic (cost of equipment) and political (security of supply).

The question is when these limits will be reached and what will be the effectiveness of energy use equipment. In all likelihood, significant progress can still be made. As it will be discussed in Chapter 9, it is on this assumption that proactive scenarios regarding the evolution of energy consumption are based, their designers believing that significant reductions in per capita energy consumption are possible while maintaining current standards of comfort.

Energy conservation appears to be the best solution for developing countries wishing to meet their energy needs without their useful primary energy consumption growing at a rate that is incompatible with their financial resources. Only energy conservation can continue to ensure the economic development of the entire planet while limiting the pressure on energy supplies and the environment.

To evaluate the potential of energy conservation measures, conservation supply curves are very interesting tools that combine technical and economic parameters. They provide (in abscissa) the cumulative energy conservation that can be achieved based on the cost of possible measures (in ordinate).

The curve in Figure 8.7, established by UNIDO [UNIDO, 2010] provides the cumulative annual primary energy conservation potential achievable by improving compressed air systems in Europe. Compressed air is indeed used in many industries or public works to provide energy to various tools or devices. These systems are clean, reliable, sometimes noisy, but rarely optimized energy-wise. This curve shows in abscissa the cumulative conservation, expressed in billion kWh per year, with, in ordinate, the average cost of each of these savings, in $/MWh. A summary of the different measures is listed in Table 8.1, their reference number being shown on the curve. The third and fourth columns of this table give the values of the curve's abscissa and ordinate, and the fifth shows the cumulative potential reduction in CO_2 emissions.

By plotting the average price of electricity on the cumulative curve, one can clearly see the separation between the economically feasible measures (1 through 9), corresponding to a potential of 18,519 GWh/year, and those that are technically feasible, but too expensive (24,857 GWh/year).

Such curves are therefore very useful methodological tools for assessing potential energy conservation in a specific area for a country or a group of countries.

A similar graph, established by McKinsey Sustainability & Resource Productivity [Kiely, 2010], illustrates the very large potential for energy savings in the United States.

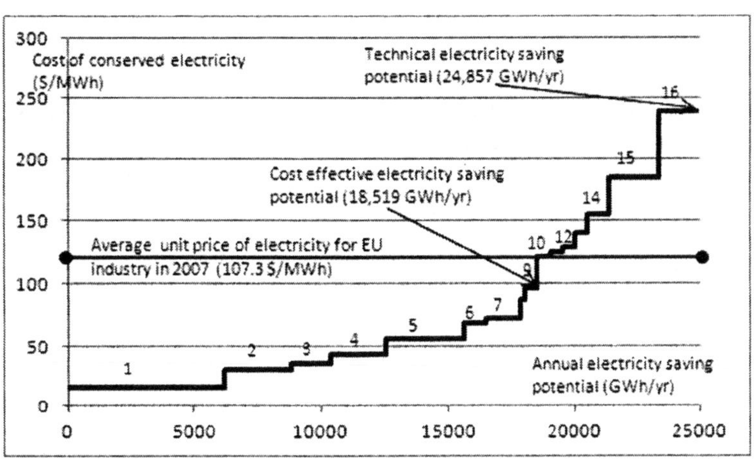

Figure 8.7 Potential electricity conservation achievable by improving compressed air systems in Europe.

Table 8.1 Cumulative electricity savings for compressed air systems in Europe.

N°	*Measure*	*Potential GWh/yr*	*Cost $/MWh*	*CO_2 emiss. reduction ktonne CO_2/yr*
1	Fix leaks, adjust compressor controls	6,190	17	2,700
2	Install sequencer	8,874	31.7	3,869
3	Initiate predictive maintenance	10,381	36.9	4,526
4	Improve end use efficiency	12,561	44.1	5,477
5	Eliminate inappropriate compr. air uses	15,606	56.3	6,804
6	Address restrictive end use drops	16,438	68.5	7,167
7	Eliminate artificial demand	17,795	73	7,759
8	Correct compressor intake problems	17,980	87.6	7,839
9	Use drains with zero loss type	18,519	96.2	8,074
10	Correct excess press. drops in main line	19,039	121.5	8,301
11	Reduce critical pressure loss	19,540	124.8	8,519
12	Install dedicated storage	20,024	129.1	8,730
13	Correct excess supply side pressure drop	20,492	139.9	8,934
14	Match air treatment to use side needs	21,381	154.9	9,322
15	Improve compr. part load (variable N)	23,366	184.7	10,188
16	Size replacement compressor	24,857	238.8	10,838

8.3.6 Carbon capture and storage

Carbon Capture and Storage (CCS) aims, as its name implies, at recovering CO_2 from combustion of carbonaceous materials (mainly fossil fuels but also biomass), injecting it underground at high-pressure in adequate geological formations, and storing it in the long term. This is a mitigation measure designed to reduce the carbon intensity of energy. It could be described as a "second order measure" since the amount of carbon is not reduced at the source, in terms of fuel, but at the output: the amount of CO_2 emitted is the same, but its release into the atmosphere is avoided [Gicquel, 2011].

One can identify two advantages of this measure:

- firstly, it is a source of flexibility in the short term. Indeed, the known fossil fuel resources represent 5 times the amount of carbon released into the atmosphere by human activities since 1860. Restrictions on the use of this fuel will not stem from any constraint on resources, whatever the uncertainties on the technological developments needed to exploit unconventional sources, and geopolitical aspects. Moreover, if any decision or political incentive to abandon this type of fuel is put in place, a "reaction time" of actors and technologies must be taken into account. The IPCC estimates that, whatever the scenarios, fossil fuels will remain the main source of energy until 2020. CCS would modulate these unavoidable emissions, and make technological adaptations cheaper: fossil fuels would still be used, but their impact lessened in terms of CO_2 emissions. This short-term benefit has been extensively studied and evaluated, especially by the IPCC [Metz et al., 2007];
- combined with other technological advances, CCS would provide a much greater potential for reducing emissions: it could allow one to produce new clean

non-fossil energy carriers from fossil fuels, such as hydrogen. Its scope would thus be greatly expanded.

An important argument in favor of CCS is that its feasibility is almost established. CO_2 emissions must first be trapped or caught in a gas stream that is often mixed (for example, the gas exiting a power plant consists of approximately 15% CO_2). Almost pure and high pressure CO_2 must then be transported to its storage location, where it must be injected into geological or marine formations with characteristics appropriate to store it in the long term.

For each of these steps, there are now several technical solutions at various degrees of technological maturity, but on the whole their feasibility has been demonstrated and they have been commercialized in other fields. Most of these solutions are tightly controlled by the oil and gas industry, since they have experience in particular in the transport of gas under pressure and expertise in geological layers. The trapping itself is the least technologically mature step of the process (at the demonstration stage only), and thus the one awaiting the most progress.

Three solutions are now envisaged to achieve this progress in trapping: post-combustion capture, pre-combustion capture, and oxycombustion, the latter being the least mature. The recovery of almost pure CO_2 produced by certain industries can be added to this list. Transportation from the production site to the storage site could be mainly done by pipeline, with possible use of tankers, but it does not pose new problems in relation to current transport of various gases and fuel products carried out by the oil and gas industry.

Geological storage could mainly take place in old oil or gas reservoirs, in saline aquifers in deep formations, or in coalfields that are difficult to exploit. These formations are indeed regarded as offering the best guarantees in terms of permeability and ability to retain and/or hold the gas under pressure. Three industrial projects already exist: Sleipner in the North Sea, Weyburn in Australia, and In Salah in Algeria, to store from 1 to 3 Mt of CO_2 per year. Pressurized CO_2 injection is used in some cases by the oil industry for enhanced oil recovery. Ocean storage, based on the principle of dissolving CO_2 in water at great depths, is currently under study. The potential seems immense, but the risks for ecosystems (as dissolved CO_2 acidifies water) and climate regulation in the long term are still largely unknown, hence the great caution. In terms of storage capacity, estimations suggest that 1,685 to 11,100 Gt of CO_2 could be stored (without ocean storage), but the various models show that this is not a limiting factor.

Before detailing the various CCS techniques and the progress that can be expected, let us point out the main obstacles still existing concerning their implementation on a large scale:

- The first limitation concerns storage, and uncertainties related to leaks. Occasional leaks, during transport for example, are a danger for the people living nearby, but not a real problem on a large scale. The difficulty comes from more diffuse leaks over a long period, once the CO_2 is stored: to what extent can we consider that CO_2 storage in a geological formation is a permanent storage, particularly in relation to the time scale of climate change?
- Environmental risks associated with CCS are also poorly understood. We know for example that some of the possible processes or technologies, such as post-

combustion capture, require the use of organic solvents: how will these solvents be produced and recycled, and in what quantities?
- Finally, many legal matters are still pending, particularly related to the right of the soil.

As mentioned above, there are three main systems for capturing CO_2: post-combustion capture, pre-combustion capture and capture by oxycombustion.

Each of these systems involves a step of gas separation. Separation processes vary depending on the trapping system, but they may share some of their separation techniques, of which there are four major families: cryogenics, use of solvents, membranes or solid adsorbents.

Post-combustion capture refers to the separation of CO_2 from gases emitted by burning fossil fuels or biomass. Instead of rejecting combustion fumes directly into the air, gases are processed to extract CO_2, which is compressed, transported and stored. The remaining gas is then released into the atmosphere. The separation technique commonly used involves chemical or physical absorption. One method announced as the most promising is the absorption in refrigerated ammonia, in which Alstom is currently heavily investing (chilled ammonia process or CAP).

Beyond the existing industrial applications, post-combustion capture could specifically be applied to existing oil, coal or gas power plants, including pulverized coal plants or natural gas combined cycles.

The process of pre-combustion capture exists, but on a small scale. It typically consists of two steps.

The first step generates a mixture of carbon monoxide (CO) and hydrogen from the primary fuel. Two reactions are possible: steam reforming and partial oxidation or gasification.

The second step is a conversion step which converts CO to CO_2 using water gas reaction.

Finally, CO_2 is extracted from the mixture CO_2/H_2, whose initial concentration is 15% to 60% and pressure 20 to 70 bar. CO_2 can then be stored. Note that pre-combustion capture entails a significant energy cost, because of the need to convert all the CO produced during the first stage into CO_2, before entering the combustion chamber. Part of the initial fuel LHV is lost in the process.

Two applications are possible for pre-combustion:
- the production of a fuel with virtually no carbon (H_2). Hydrogen need not be pure, but the less carbon the fuel product contains, the less CO_2 its combustion produces;
- a decrease in the carbon content of fuels. For example, coal has a low H/C ratio. Using pre-combustion, one can produce a fuel with a larger ratio and capture the corresponding carbon released as CO_2. The subsequent use of the obtained fuel releases far less CO_2.

In the oxyfuel capture technology, combustion is realized in pure oxygen or in a mixture of oxygen and recycled gas rich in CO_2 or water. Thus, the gases produced by the combustion are composed mainly of water, CO_2 and excess oxygen for complete combustion. In order to implement such a combustion, oxygen must be separated from nitrogen, the main component of air, which usually has a significant energy cost.

CCS is an interesting option for mitigation. It should be attainable at a cost deemed "economically and socially acceptable", since it does not impact consumption levels. However, its cost is not zero, because it leads to additional energy consumption. CO_2 capture would be especially relevant for large emitters such as power plants using coal or gas and for certain industries such as iron, steel, cement or heavy chemicals. It is much more expensive and difficult to extract CO_2 from small or mobile sources.

8.4 A CHRONICALLY UNSTABLE OIL MARKET

Another important recent development is related to the supply and demand for oil.

We saw in Chapter 5 that the main objectives of energy policy of energy-importing countries are generally threefold:

- reduce the energy content of their GDP;
- increase their domestic supply of energy;
- diversify their energy supplies.

These three objectives have been implemented to varying degrees by every oil-importing country, sometimes in a context of declining GDP growth or even recession (for the poorest countries).

The measures undertaken to achieve these goals have not paid off immediately, but the effects began to be apparent in the early 1980s.

The energy supply has greatly increased due to the emergence of new producers wishing to take advantage of high oil prices to develop alternative national energy sources (coal, gas, nuclear, renewable), wherever possible.

The global energy demand fell in response to crises, structural changes following conversions towards less energy intensive industries, and finally to successes in energy conservation.

The diversification of supplies has resulted in:

- firstly, diverting oil importing countries from oil, which have sought substitutes whenever possible, even imported ones, which explains the growth of coal and natural gas markets despite the partnership constraints associated with such fuels, even at the cost of a small overhead;
- secondly, leading importing countries to diversify the origin of their supplies to avoid being in a situation of political dependence *vis-à-vis* a small group of countries.

8.4.1 Consequences on the demand to OPEC

This resulted in the demand to OPEC falling sharply, before picking up after 1985, as shown in the graph in Figure 8.8 (left axis in Mt, and right axis in Mb/d).

Energy exchanges have moved from a situation where a producer cartel could impose its prices by lightly reducing its production so that demand became superior to supply without alternative in the short term for importers, to a situation where supply exceeded demand so that the same cartel saw its production almost halved in volume in less than 15 years without being able to curb the fall nor regain its market share.

However, since 1985, the growth in demand has again been mainly supplied by the OPEC countries, as these countries have resources that are easier to technically

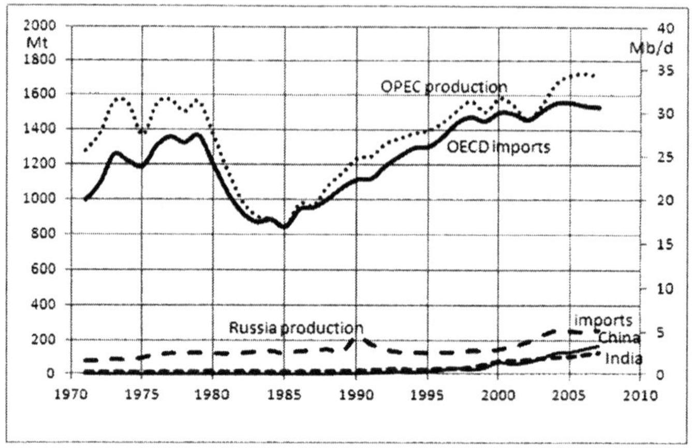

Figure 8.8 Production and imports of oil.

and economically exploit. Production areas of other groups of countries are in fact more difficult to value.

As shown in Figure 8.9, since about 2002 and despite increasing oil prices, the non-OPEC oil production has begun to stabilize, whereas it had increased by about 35 Mt/year during the previous decade.

This stabilization comes from the depletion of some deposits because of their low oil capacity.

8.4.2 The peak oil debate

This point is well illustrated in Figure 8.10, which shows that after hitting a maximum in 2000, oil production from the North Sea is now sharply decreasing.

Akin to what is happening in the North Sea, the production profile of all oil wells is a bell curve, showing a maximum: the so-called peak. This profile is explained by the different operating phases of the basin, occurring over a period of several decades:

- commencing operations, represented by the left side of the curve, lasts a few years before the maximum capacity is reached;
- maximum production, corresponding to the peak;
- decrease in production, at an average rate of 4% per annum, depending on geological conditions.

Departing from this familiar graph depicting the life of a reservoir, various experts have sought to represent the global oil production in a similar way, assuming that it would also follow a bell curve, the knowledge of the peak value and the period in which it would occur being of great interest for analyzing the oil issue according to them.

The first geologist to have done this in the 1950s was Marion King Hubbert, who predicted in his scenarios the peak of U.S. production, actually observed in 1970.

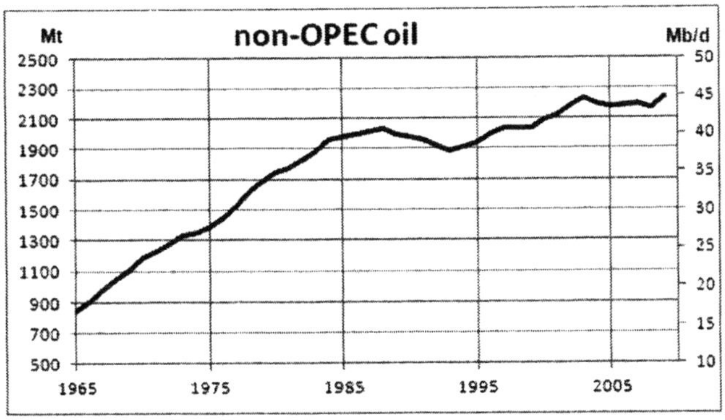

Figure 8.9 Production of non-OPEC oil.

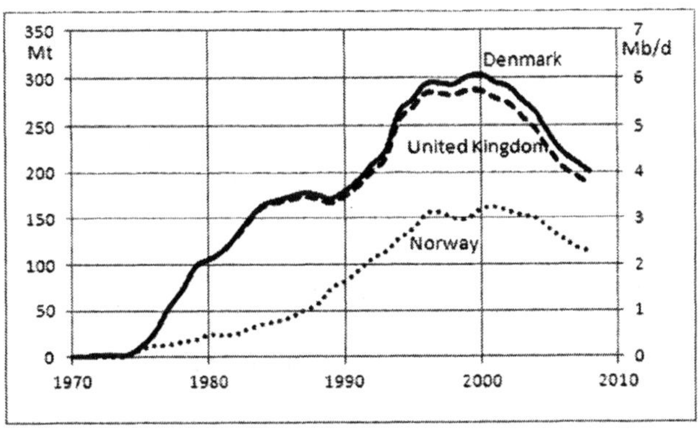

Figure 8.10 Production of oil in the North Sea.

Since then, the debate is regularly publicly conducted between "optimists" and "pessimists", the former arguing that technological advances will push back the arrival of the global peak for a long time, while the latter, grouped within the ASPO (Association for the Study of Peak Oil and Gas), consider it unlikely that large deposits will be discovered and, as the overall production rate is increasing, predict that the peak could occur in the near future (2015).

We mention this debate for the reader's information, as the oil issue cannot be reduced to an analysis of the capacity and production volumes, given its geopolitical dimension.

Some experts consider it very likely that global oil production will not follow a simple bell curve, but will instead stabilize (with oscillations) at a roughly constant level for several years or decades, between 90 and 100 Mb/d, corresponding more or less (beginning of dotted curve in Figure 8.11) to the current production (2012), before decreasing.

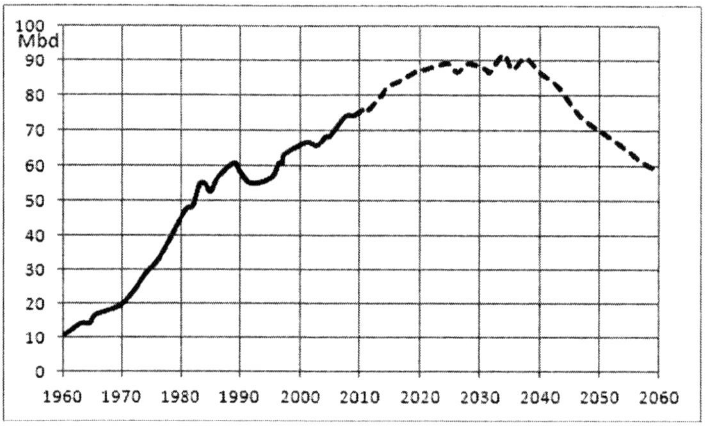

Figure 8.11 Forecast of peak oil.

8.4.3 Interaction between oil demand and price

Oil demand appears to be a direct function of its price, probably because it is one of the most versatile and least expensive (investment-wise) forms of energy, at least in terms of use.

When the price is high, oil-consuming countries turn away from it and undergo austere energy policies, which bear fruit but are challenged when prices fall. Economically, there is no doubt that these behaviors are justified, but they result in favoring short-term policies that are quite detrimental to the preparation of the needed long-term transition.

The graph in Figure 8.12 shows the changes in demand and oil prices (in 1983 dollars), reduced to an index of 100 in 1965. It shows that the declines in demand are always secondary to significant price increases (1973, 1979, 1998, 2008).

For example, the threshold for competition between heavy fuel oil and coal for electricity generation is around an oil price of $15–18 per barrel, and may be revised upwards due to possible environmental protection measures. Many coal plants in operation being also equipped to operate on oil, one cannot dismiss the possibility of a significant increase in oil energy intensity if its price falls below a certain value.

Low oil prices additionally slow down the production of non-OPEC oil. Below $15 a barrel, small marginal fields would gradually close. Companies, not releasing any more profit, would be unable to continue significant exploration campaigns, so that the reserve replacement would no longer be assured.

The market has a natural tendency to capsize, the supply being unable to follow. Instability seems to be the natural behavior of the global oil market.

Without doubt, it is now time for the international community to attempt to regulate as much as possible the oil market. On the economic front, it seems that everyone can benefit from it. On the political level, however, OPEC can benefit from instability, since it can influence the fate of the world when the context is favorable.

Given the cost structure of oil, OPEC indeed has every opportunity to advocate lower prices without significantly losing money on a regular basis. It can and will

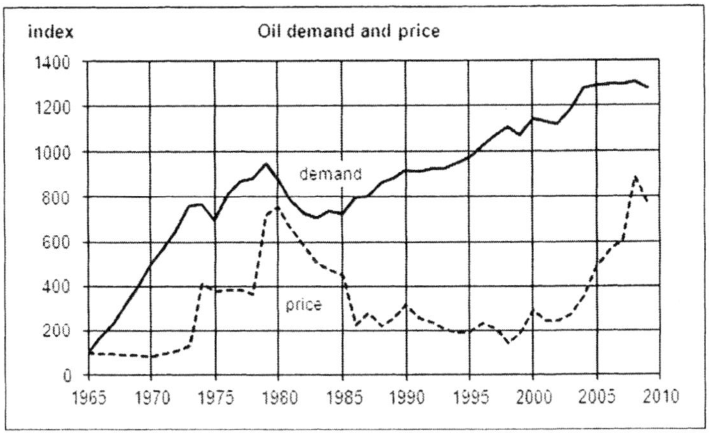

Figure 8.12 Oil demand and prices.

increasingly regain its market share by driving competitors if not out of business then at least into serious financial difficulties.

REFERENCES

Gicquel R. Energy Systems: A New Approach to Engineering Thermodynamics, January 2012, CRC Press, ISBN-13: 978-0415685009.

Kiely T., Energy efficiency, A Compelling Global Resource, McKinsey Sustainability & Resource Productivity, 2010.

MEDDE, Repères, Chiffres clés du climat France et Monde, Paris, 2013.

Metz O.R., Davidson P.R., Bosch R., Dave L.A., Meyer (eds), IPCC (2007). Climate Change 2007: Mitigation of Climate Change. Contribution of Working Group III to the Fourth Assessment Report of the Intergovernmental Panel on Climate Change, Cambridge University Press, Cambridge, United Kingdom and New York, NY, USA.

UNIDO, Motor Systems Efficiency Supply Curves, December 2010, Vienna.

FURTHER READING

Balaceanu J.C., Développement Technologique et Prospective Énergétique: les scénarios technologiques, Annales des Mines, Paris, janvier 1990.

Bourgeois B., Finon D., Martin J.M., Ed. Énergie et changement technologique, une approche évolutionniste, Economica, Paris, 2000.

Brown M., Sovacool B., Climate Change and Global Energy Security, The MIT Press, 2011.

Chapter 9

Outlook

9.1 PRESENTATION

Throughout this book, we analyzed many aspects of the energy issues which allowed us to better understand the current situation. Now comes the natural question: what could our energy future be? It arises all the more acutely as our assessment showed the existence of a number of threats, such as the depletion of fossil fuels (first oil, then natural gas), the risk of future oil shocks, and the increase of the greenhouse effect.

In this context, this chapter will start by examining what the prospects for energy consumption and supply look like in the long term (50–100 years) as well as in the medium term by analyzing several scenarios, after which we will try to draw a few conclusions on the main perspectives.

9.2 ENERGY SCENARIOS

To try and anticipate future trends, it is interesting to sketch scenarios regarding growth of energy consumption and oil supply. Given the didactic purpose of this book, we will begin by giving some methodological guidance, before comparing a few scenarios.

Prospective methods differ from forecast methods in that their scenarios do not seek to predict in any way what the future will be and therefore do not try to be prescriptive. They are simply exploration tools aimed at getting an idea of what the sets of considered assumptions imply. Indeed, it is often possible to project the possible (or desirable) evolution of some key variables, either because they roughly follow the laws of statistics (populations) or because they may correspond to a proactive policy project (GDP growth rate).

The value of these scenarios, as we shall see in this section, is that they highlight key trends and thus define a useful framework for thinking about the future.

9.2.1 Long-term scenarios (2100)

In this section, we will review long-term scenarios that we initially presented in the first edition of this book published in 1992. The comparison with the real evolution of consumption since will indeed prove interesting.

The scenario method, as explained above, is exploratory rather than normative. To explore the future, it is necessary to extrapolate the present situation, by minimizing uncertainty. To do so one has to try and link the studied variables to stable indicators, if available. As extrapolation is always risky, the choice of indicators is of fundamental importance.

To project into the future, one might intuitively think of using the growth rate of the studied variables. However, this method, which translates mathematically into extending a curve by its first derivative, is valid only in the vicinity of the starting point. Beyond that, it can lead to absurd results, especially if the curve is not monotonic.

If one is interested in what the energy consumption could be in the very long term, it is therefore out of the question to extrapolate current average growth rates, because they will certainly not be valid in a few years' time, as the situation will then likely be very different from the present.

9.2.1.1 Assumptions on population and per capita consumption

It is therefore necessary to use a more stable intermediate value basis for our extrapolations. On several occasions throughout this book, we expressed the total consumption as the product of the population by the per capita consumption. It turns out that the evolution of populations is a rather stable phenomenon, which is the subject of numerous studies. We can thus rely on the existing prospective studies in this area, and, in order to determine the energy consumption, set additional assumptions on the level of per capita energy consumption.

The evolution of the world's population in recent centuries is precisely known (Figure 3.1).

Fertility generally drops down to a certain threshold when the per capita standard of living increases, so that it can be expected that over time the global population will cease to grow exponentially and will stabilize at a certain level, called the hypothetical stationary population by specialists. Besides fertility, it is understood that many other limiting parameters exist, in particular related to the possibility of providing acceptable living conditions on Earth for all humankind.

The World Bank estimates that the global hypothetical stationary population could be around 12 billion, and could be reached in the course of the 22nd century. Note that this figure corresponds to twice the population of 1999.

Following this assumption and taking into account the very different dynamics in developing countries and in industrialized countries, we can set up a model for the evolution of the population of the type presented in Figure 9.1.

The estimation of per capita energy consumption can also be modeled. In 1988, the world average per capita energy consumption was equal to 1.6 toe/capita/year, but this figure masked wide disparities: 0.5 toe/capita/year in developing countries and more than 5 toe/capita/year in industrialized countries (Figure 6.3).

How are these values likely to evolve in the future?

Clearly, per capita consumption in developing countries should increase for these countries to have a chance to develop. However, we have seen that there is some flexibility in the industrialized countries, where it is possible to reduce the energy intensity without compromising economic growth.

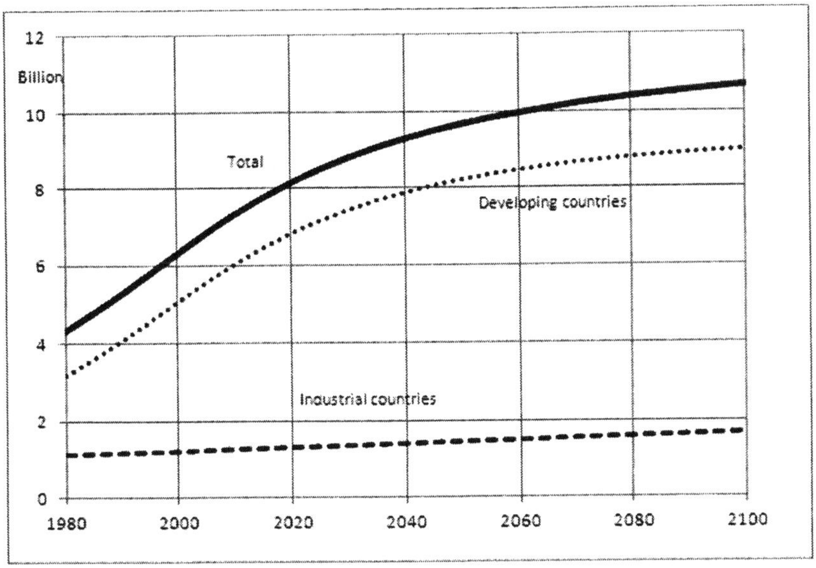

Figure 9.1 Scenarios for the world's population.

Since 1985, a large number of scenarios have been proposed to estimate how the world's energy consumption could evolve. They differ fundamentally according to the assumptions they hold about the evolution of per capita consumption.

These scenarios can be grouped into two broad categories: trend scenarios, which extrapolate recent trends in per capita consumption, and proactive scenarios, which require a break with past trends, justified by the political will to redress what is considered an undesirable trend.

The problem is that the outcomes of these two approaches are very different, if not contradictory. The degree of uncertainty is such that we cannot be too assertive on the likely evolution of the global energy consumption.

To illustrate the achieved results, we included in the following study the assumptions of a number of scenarios published in the 1980s, and we compared estimates of global energy consumption to which they lead. For the comparison to be meaningful, we have retained the same pattern of population growth, drawn from World Bank data, which sometimes differs from that which had been retained by the authors of the energy scenarios. Besides, where the changes in per capita consumption were not indicated, we interpolated them between the limit values. Some scenarios were established over a limited time horizon (2020). We took the liberty to extrapolate them until 2100 and indicated when relevant the assumptions upon which such extrapolations were based.

The selected scenarios are:

- The Goldemberg scenario [Goldemberg et al., 1988], very proactive, based on the idea that technological breakthroughs could lead on the one hand to a large reduction in per capita consumption in industrialized countries (from over 5 toe/capita to about 2 toe/capita in 2025), and on the other hand to stabilize per capita consumption in developing countries at around 1 toe/capita, while ensuring from that date a comfort level roughly equivalent to that of Western Europe in the late 1970s. This scenario being limited to 2020, we extended it with the assumption that per capita consumption levels remained stable after 2020 in developing countries, and continued to decrease to 1 toe/capita in industrialized countries;

- Two proactive scenarios, "Jeremie" and "Noé" (*Nouvelles options énergétiques* in French or New energy options) established by B. Dessus and F. Pharabod [Dessus & Pharabod, 1990] using an approach inspired by the Goldemberg scenario, but a little less optimistic about the possible pace of introduction of new efficient technologies, especially in developing countries, which led the authors to push back to 2060 and 2100 instead of 2020 the dates of stabilization of per capita consumption at the level proposed by Goldemberg;
- A trend scenario, as the one designed by the World Energy Conference, called "WEC low" [CME, 1989], which implied a stabilization of per capita consumption in industrialized countries in 1990 (5.5 toe/capita), and a regular increase of per capita consumption in developing countries. This scenario being limited to 2020, we extended it assuming a continuation of trends;
- A limit scenario, whereby it is assumed that the per capita consumption in developing countries and in industrialized countries would stabilize in 2100 to a common value equal to 3.5 toe/cap. We will explain further wherein its value lies.

The Goldemberg scenario, published by its authors in 1986, had a large impact because it departed radically from conventional scenarios published beforehand. It still deserves to be commented upon.

It is based on the idea that the continued increase of energy consumption at the same pace as in the past is no longer possible, given its overall prohibitive cost. Due to the growth prospects of the population in developing countries, it seems impossible that these countries can ever achieve per capita consumption similar to those of European countries in the early 1980s, especially as such a growth of consumption would result in a major worsening of the firewood and emission crises entailing, from this point of view, significant threats to the greenhouse effect and reprocessing of nuclear waste.

How can this conundrum be solved? The basic idea is that technological advances in equipment efficiency have been made since the first oil shock. If we could generalize the use of the best technologies, all specific consumptions would drop drastically, and the problem would have completely changed.

In fact, according to the authors, it would be possible, by 2020, to ensure that people in developing countries attain a rather acceptable level of energy comfort while limiting consumption to 1 toe/capita. Meanwhile, in industrialized countries, a value of 2 toe/capita could be achieved (dividing by 2.5 the value of 1990).

We believe that four fundamental objections can however be made to the authors' model:

- the values used for some specific consumptions correspond to experiments carried out on a small scale in very specific conditions that cannot be extrapolated without great precautions. This is the case particularly for energy consumption used for housing in Europe, based on some Swedish achievements that were not

transposable in general, if only because of the diversity of architectural patterns and lifestyles;
- the rate of diffusion of new technologies, whatever their qualities, is limited by the renewal time of the existing fleet;
- the authors extrapolated their conclusions to all developing countries although they were established for a standardized country. We already emphasized the diversity of energy situations in developing countries and the fragility of any exercise that would consider them homogeneous;
- how the transfer (to developing countries for example) of efficient energy technologies would be carried out is not explained, although it is a major unsolved problem, despite the efforts UNCTAD has been making for decades.

However, despite these limitations, the work done by Goldemberg and his co-authors had the merit of calling out to energy specialists and getting them to think more about future developments. The Jeremie and Noé scenarios appear as an effort to strike a more realistic note than the previous scenario. Being aware that 2020 is at hand and given the long lead times for implementation of energy projects, the authors have added 40 to 80 years to the limits chosen by Goldemberg.

They retained, however, Goldemberg's concern to achieve a "sustainable world", wherein a balance would be found between energy consumption and conservation of resources, growth in developing countries and the global environment.

Including changes in the concentration of carbon dioxide in the atmosphere and the accumulation of long life nuclear waste in their analyses, Jeremie and Noé show that it is possible to limit them, but only at the cost of a deliberate policy that contrasts with previous trends.

The other two scenarios allow us to see precisely how these trends differ from the previous ones. The first one (WEC low) already corresponds to a stabilization of consumption in industrialized countries, and a slow down in the growth of consumption in developing countries.

The limit scenario shows the implications of a more gradual reduction of consumption in industrialized countries, with a proactive policy in developing countries and technological advances not being made as quickly as hoped by Goldemberg.

The graphs in Figures 9.2 and 9.3 summarize the assumptions used with respect to the evolution of per capita energy consumption.

9.2.1.2 Comparison of long-term scenarios

As shown in the graphs in Figures 9.4 and 9.5, which only differ in their projection horizon, the outcomes of these assumptions are very different, which leaves a lot of uncertainty about the value of the global energy consumption in the future.

The differences are considerable: in 2020, the estimated global consumption varies in the range of 1 to 2 depending on the scenario.

By 2100, this range reaches 1 to 3. Clearly, the Goldemberg, Jeremie and Noé scenarios belong to the same family and differ from each other only in the second order.

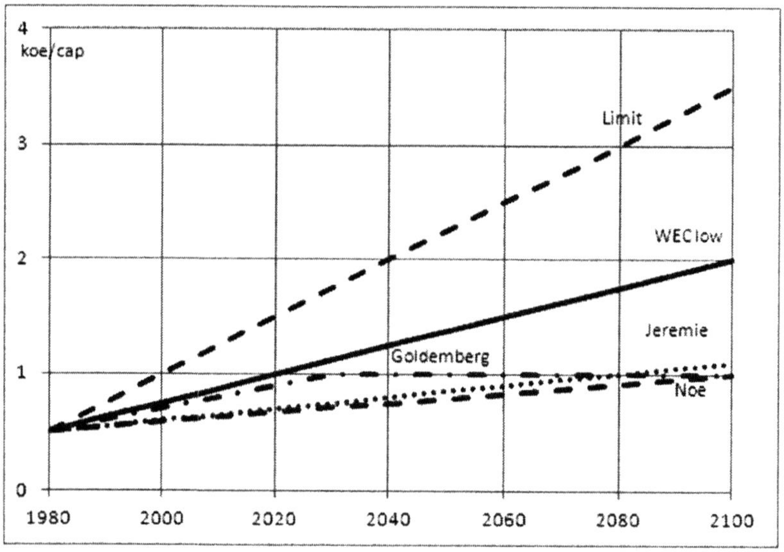

Figure 9.2 Energy consumption assumptions (developing countries).

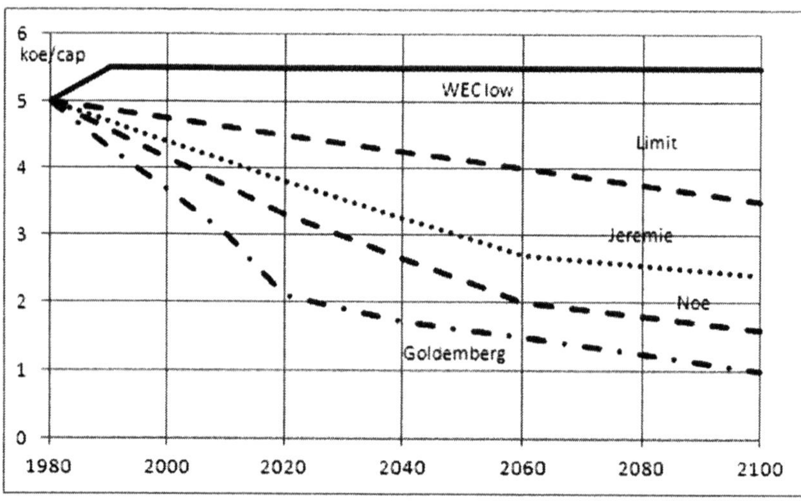

Figure 9.3 Energy consumption assumptions (industrialized countries).

9.2.1.3 Comparison with actual changes

The comparison of these scenarios, prepared in 1990, with actual developments is given in Figure 9.6, the latter appearing in dashed lines. They lie between the WEC low and Jeremie scenarios, the passage from one to the other having occurred around 2000.

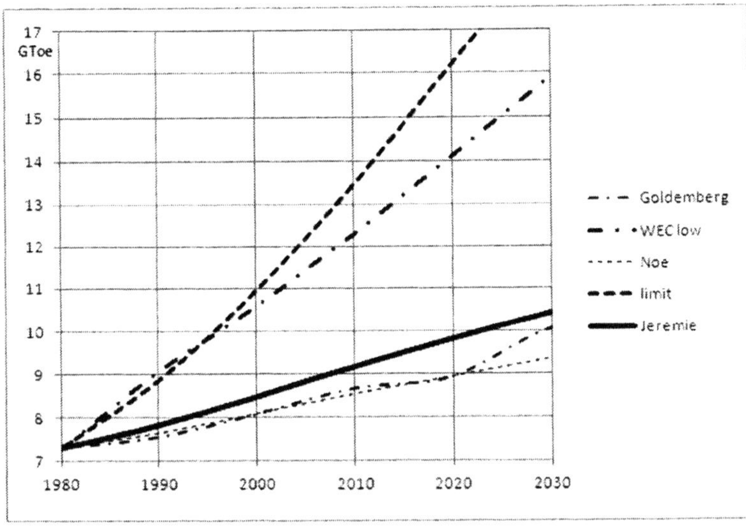

Figure 9.4 Comparison of global energy consumption in the various scenarios.

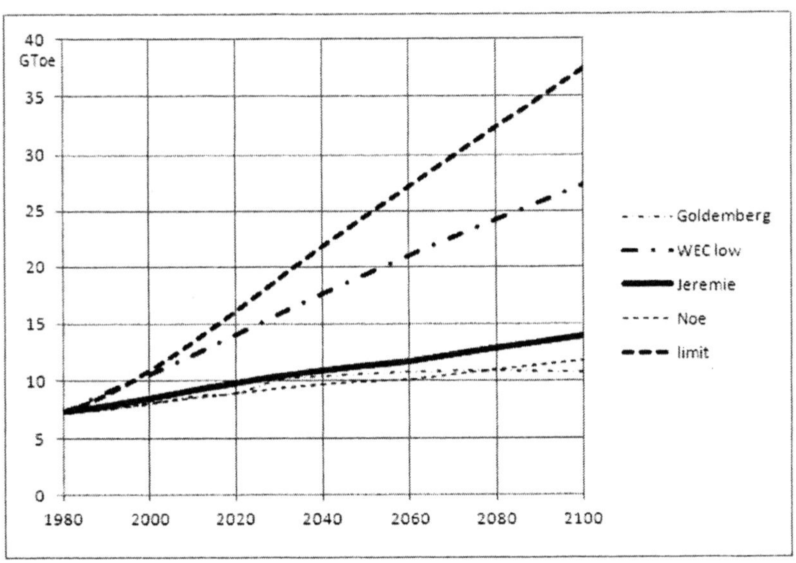

Figure 9.5 Comparison of various scenarios.

The sharp increase in global energy consumption at this time was due to the economic growth of a number of developing and emerging countries.

This comparison highlights the robustness of the assumptions on which the long-term scenarios were built, although it is clear that the range of values they yield is very broad.

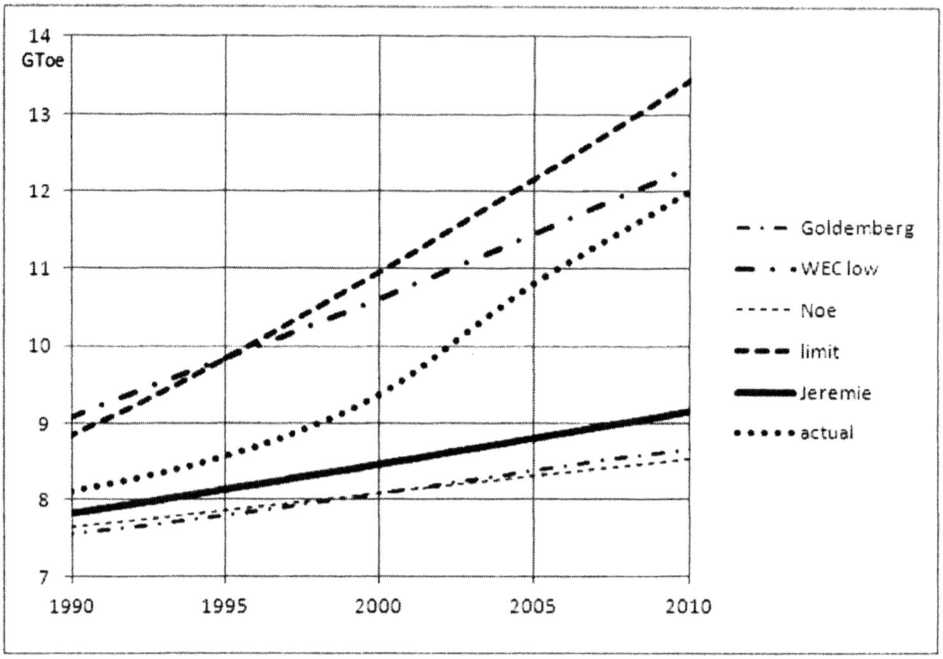

Figure 9.6 Comparison with actual changes.

Let us already indicate that most of the medium-term scenarios currently studied and presented further provide energy consumption values for 2030 that fall between scenarios WEC low and limit.

9.2.1.4 The available energy

The limit scenario has been developed to estimate, even in cases of a strong growth in consumption, what lead time the fossil fuel reserves would provide us to prepare a transition to longer-term renewable energy such as solar energy or nuclear fusion.

On the basis of the limit scenario, it is possible to calculate the cumulative consumption and hence the amount of energy it would take to implement it (Figure 9.7).

Cumulative consumption increases very rapidly, reaching about 500 Gtoe in 2020, 1,000 Gtoe in 2045, and 1,400 Gtoe in 2065.

In 1990, fossil fuels accounted for nearly 90% of the global commercial energy balance, and their proved reserves were estimated at about 750 Gtoe, 100 Gt of oil, 100 Gtoe of natural gas and 550 Gtoe of coal.

Knowing that the reserve concept is inherently dynamic, one is entitled to assume that, because of future discoveries and advances that will be made with respect to the rate of oil recovery in the following decades, proved reserves could correspond to a cumulative production of about 1,000 Gtoe by 2065.

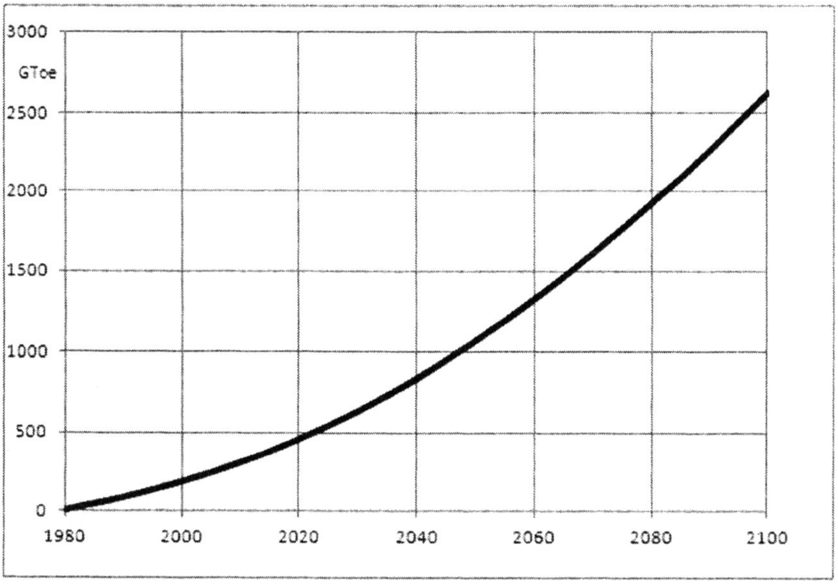

Figure 9.7 Cumulative energy consumption.

In addition, non-conventional fuels represent a considerable potential in the future, be they extracted from non-conventional oils (tar sands, heavy oil and extra-heavy oil shale), coal (gasification or liquefaction) or biomass.

Cumulative consumption figures show that, even within a scenario with such a high consumption, conventional hydrocarbons only would suffice to address all energy needs until 2045, that is to say for 55 years, the scenario origin being 1990. At that date, the contribution of non-conventional fuels could already be significant, providing perhaps up to 400 Gtoe, which would give another 20 years of respite on the basis of fossil fuels alone.

However, 70 or 80 years represent a sufficient amount of time for a radical shift in energy balance to take place. Renewable energy such as solar energy or fusion could pick up the baton, provided that the right incentive policies are implemented quickly.

Moreover, since the penetration of renewable energy sources has already begun and will be gradual, we can conclude that, even in a scenario of high growth for both the world population and per capita energy consumption, reserves of fossil fuels across the planet should be sufficient for mankind to make its transition to the energy sources of the future, renewable by necessity.

This scenario also shows the strategic role that non-conventional fuels will play in the future. To illustrate this, we present in Figure 9.8 a graph showing from which energy sources the liquid fuel supply could come, as estimated by BP in 2012 in its Energy Outlook 2030 [BP, 2012].

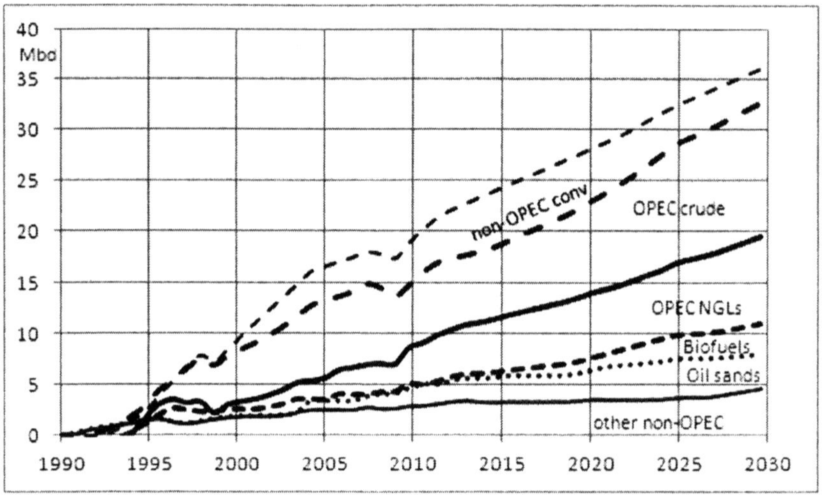

Figure 9.8 BP Outlook 2030 liquid supply growth.

The rapid penetration of unconventional fuels is obvious: oil sands, biofuels and Natural Gas Liquids (NGLs) will be developed, the total of which will represent in 2030 an additional capacity of about 20 Mb/day, i.e. 5 more than conventional crude.

9.2.2 Medium term scenarios (2030–2050)

The main value of long-term scenarios is to put into perspective the need for energy induced by economic development and energy resources available in the medium and long term.

Medium-term scenarios allow us to perform much closer projections and can be used to guide the energy policies of countries and groups of countries. These projections are of much more immediate interest than the long term ones, and various organizations regularly publish such scenarios.

These scenarios are often based on a sound extrapolation of the trends in the economic growth of the major countries of the world, and evolution of energy intensity, by applying the equation:

$$C_{ENER} = \text{GDP } C_{ENER}/\text{GDP} = \text{GDP EI}$$

In section 3.6.6 we saw that a convergence of energy intensity values can be observed in many countries, with values ranging between 0.1 and 0.2 koe/\$2010, with a general downward trend.

The scenarios that we will discuss here are those published by the International Energy Agency in its World Energy Outlook [IEA World Energy Outlook, 2008],

BP (BP Energy Outlook, 2030) [BP, 2030], and Shell (Shell energy scenarios to 2050) [van der Veer et al., 2008].

The IEA offers three scenarios (current policy, new policy, and 450):

- the first is the continuation of policies already formally adopted and implemented;
- the second assumes the implementation of new measures to meet national emission reduction of greenhouse gas emissions and, in some countries, plans to eliminate fossil fuel subsidies;
- the third involves energy choices to limit in 2100 the increase in global average temperature to 2 °C, which would require to limit the concentration of greenhouse gases in the atmosphere to about 450 ppm CO_2 equivalent, hence its name.

The scenario developed by BP corresponds to trend estimates based on likely developments in the global economy, current policies and technology [BP Energy Outlook, 2012].

The two Shell scenarios are called "Scramble" and "Blueprints":

- in Scramble, policymakers pay little attention to a more efficient use of energy until tensions concerning supplies appear. Similarly, emissions of greenhouse gas emissions are not seriously addressed as long as there are no major climatic changes;
- in Blueprints, countries tackle proactively their economic development, energy security and environmental pollution. The carbon economy grows, which encourages the development of clean energy technologies such as carbon capture and storage of CO_2, and promotes energy efficiency measures.

9.2.2.1 Assumptions

BP's assumption for global economic growth is 3.7% per year over the next 20 years (against 3.2% between 1990 and 2010), and IEA's is lower (2.65% per year). Shell's assumption, though unspecified, is probably close to 3% per year.

Assumptions regarding the reduction of energy intensities vary depending on the scenario.

BP estimates the decrease rate of energy intensities to be 2% per year for its model, while it was only 1.2% per year over the last 20 years. For IEA, they are respectively 1.2%, 1.5% and 1.9% per year for each scenario (Figure 9.9). For Shell, it can be estimated to be 1.5% and 1.77% per annum respectively for the Scramble and Blueprints scenarios.

9.2.2.2 Comparison of scenarios

Figure 9.10 shows the results obtained for these different scenarios in terms of global energy consumption.

Only the 450 IEA scenario differs from the others, whose forecasts are very similar, although the methods of calculation, including disaggregation by region, type of use and type of energy are very different.

Note that the dynamics of the various groups of countries are very different (data from BP scenario):

- most of the growth in demand comes from non-OECD countries (Figure 9.11);

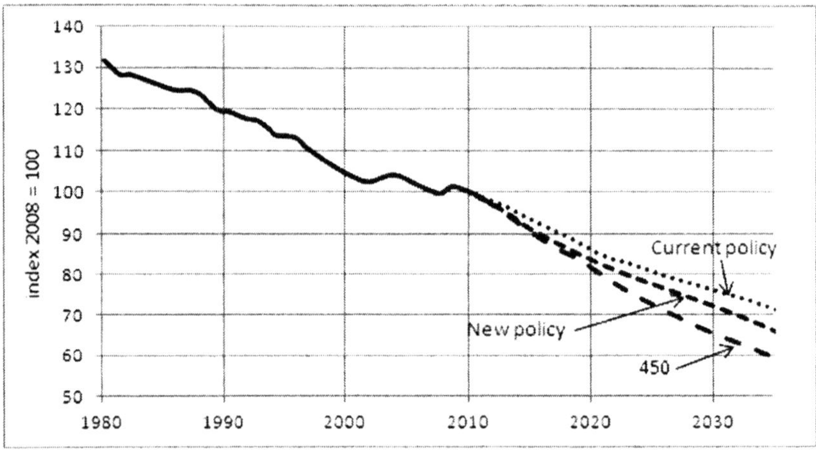

Figure 9.9 Energy intensity for the IEA scenarios.

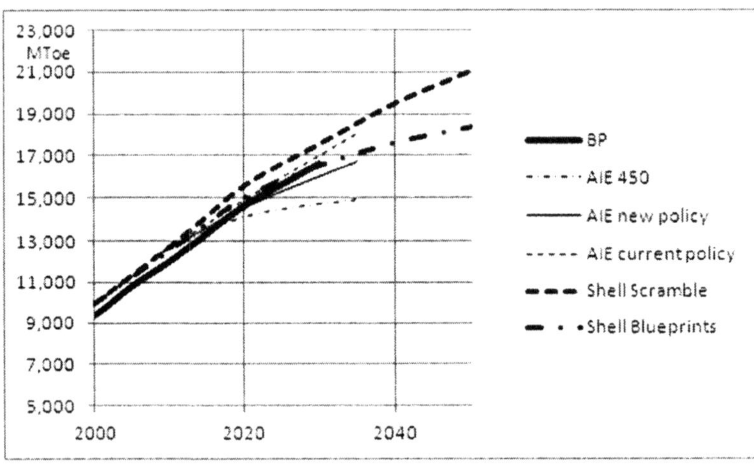

Figure 9.10 World energy consumption.

- the distribution of consumption by energy changes little, because of the projection horizon chosen (Figure 9.12). We note, however, that the three main forms of energy converge to a common value slightly higher than 4500 Mtoe;
- "liquid" energy includes not only conventional oil, but also non-conventional oil produced from oil shale, liquids from natural gas, and first and second generation biofuels, this trend reflecting the diversification of supplies.

Figure 9.13 compares these scenarios with the long-term scenarios presented before: they are close to the WEC low scenario.

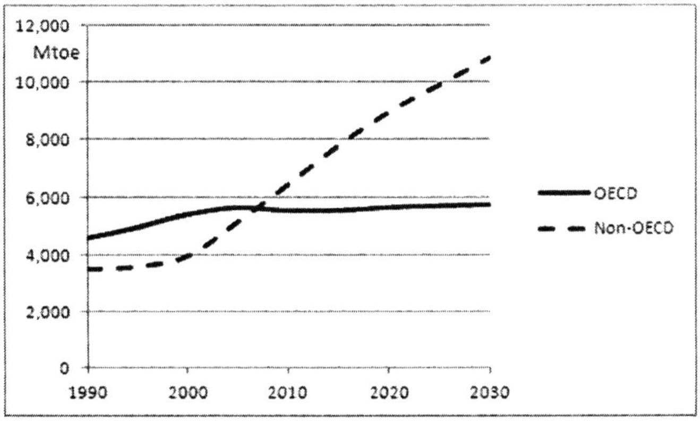

Figure 9.11 Consumption by country group.

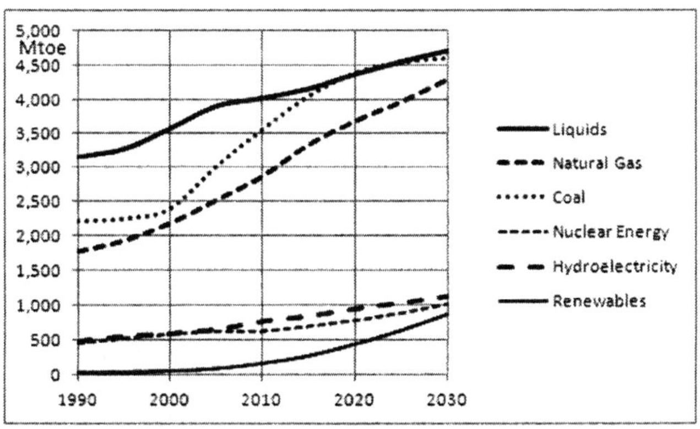

Figure 9.12 Consumption by energy source.

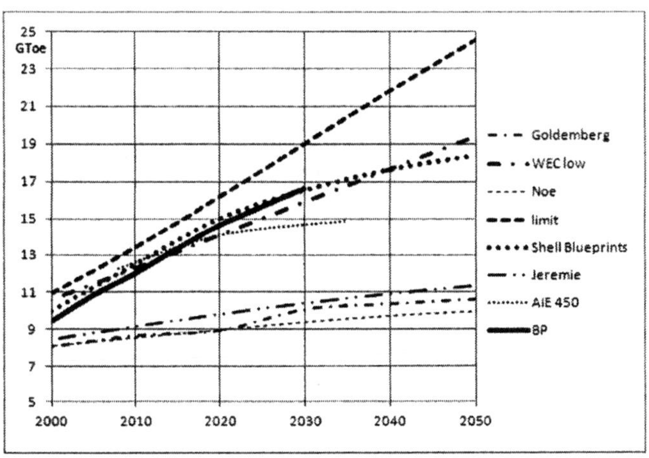

Figure 9.13 Comparison with long-term scenarios.

9.3 MAJOR TRENDS

9.3.1 A sharp increase in energy consumption

In the past, energy has led to massive substitution of human or animal labor by machines capable of performing more complex tasks, due to the development of robotics. Moreover, energy has greatly contributed to improving the comfort and well-being of mankind, by allowing the satisfaction of multiple social needs (hot water, heating, lighting, refrigeration, power supply, individual mobility, etc.).

This change undoubtedly caused a rise of standard of living and of social and economic development, and almost no one questions its advantages. However, not everyone in the modern world enjoys these benefits and the international community is unanimous in recognizing the need to disseminate them as widely as possible.

This requires rebalancing the very large difference in energy consumption between the richest countries and the poorest ones. Today, we are witnessing a change in the structure of economic production and a reduction in specific energy consumption leading to a slowdown in the growth of energy consumption in industrialized countries, hence rebalancing consumption worldwide. The process is slow, but clearly observable, and it is expected to continue in the future, especially if energy is expensive. In contrast, the consumption levels of some developing countries are so low that it should be out of the question to try and limit their growth.

This means that the level of satisfaction of the energy needs of the poor should increase significantly. As they are the most numerous people, achieving this goal is likely to involve strong increases in primary energy consumption in the future, and all the more so as the world population will continue to grow in the coming decades.

Scenarios developed by experts to estimate the future energy needs of the world differ in their assumptions about the rate of diffusion of energy efficient technology, and therefore the levels of per capita consumption in the various categories of countries. They all agree, however, on the fact that global energy consumption will continue to rise strongly in the coming decades, up to 17 Gtoe in 2030 and between 18 and 21 Gtoe in 2050.

9.3.2 A growing levy on world wealth

Beyond the concern that emissions attributable to the energy sector may have an impact on the environment, the consumption of such gigantic amounts of energy will induce considerable financial difficulties. The main reason is that alternative energy sources (likely to take over from conventional energy) will probably be much more expensive than in the past in both investment and costs, around 7 or even 20-fold. Under such conditions, the growth of the investment amount per produced toe may exceed the possible reductions in energy intensity, which will cause the share of the energy sector in the Gross World Product (GWP) to become much larger than what has been experienced so far.

In 1965, it was estimated that, at international prices, the share of the energy sector was approximately 6.2% of GWP (2% for investments and 4.2% for consumption). In 1990, it was estimated that the share of investment was 4–5% of GWP, and that of consumption 5% to 7%, depending on the price of oil, for a total of 9% to 12%.

Table 9.1 Global energy investment amount 2001–2030.

Billion dollars	*OECD*	*Non-OECD for OCDE*	*Non-OECD own needs*	*Total*	*%*
Coal	147	31	220	393	2,4
Oil	892	1 001	1 203	3 096	18,8
Gas	1 523	646	976	3 145	19,1
Electricity	4 036	–	5 806	9 841	59,7
Total	6 598	1 678	8 205	16 481	100

Table 9.2 Distribution of global energy investments 2001–2030.

%	*OECD*	*Non-OECD for OCDE*	*Non-OECD own needs*	*Total*
Coal	37	8	55	100
Oil	29	32	39	100
Gas	48	21	31	100
Electricity	41	–	59	100
Total	40	10	50	100

This increase signaled the end of the cheap energy era:

- deposits at low cost have long been exploited, the new being more difficult to access;
- environmental constraints are more severe, resulting in more complex and therefore more expensive equipment;
- electricity, which is capital-intensive, is experiencing high growth rates, as well as natural gas, which also requires costly infrastructure.

This financial constraint is not essential in itself but requires greater efforts than in the past.

Previous analyses clearly show that worldwide and in the long term, a limitation of resources should be of lesser concern than financial and environmental constraints.

In 2004, the IEA estimated that the total investment required for the global energy supply infrastructure between 2001 and 2030 would reach $16,000 billion in order to increase capacity and replace facilities which would reach the end of their use life or become obsolete during this period.

Tables 9.1 and 9.2 detail these needs, the first indicating the amounts required, and the second their distribution.

Nearly 60% of the total concerns electricity, oil and gas, each representing approximately 19%.

9.3.3 Adequate resources at the scale of the planet

The global energy consumption already exceeds 12 Gtoe, and we are entitled to question the physical availability of resources. However, as we have shown in section 9.3.1, the risk of a physical shortage appears low.

The presented calculations are of course very sketchy and questionable, but they give the order of magnitude of the minimum time lead that humanity has, in terms of fossil fuel reserves, to meet its future needs. If, as is likely, the levels of consumption set in the limit scenario are not met, the time available to prepare the "post fossil fuels" era will be longer.

However, in terms of energy, the 70-80 years figure given in section 9.2.4.1 is a long enough time to make a transition to a global mix structure that will be radically different from what we know today. 75 years from now, the production of non-conventional fuels should be well-controlled and renewable energy from direct or indirect solar energy or from nuclear fusion could begin to address a significant proportion of needs. The future of conventional energy should then be ensured.

However, it is clear that in the medium term, before the era of large-scale renewable energy, the global energy mix can only be completed by unconventional fuels: oil shale, tar sands, heavy oil and coal liquefaction. The development of these techniques is therefore strategic and poses especially difficult problems in terms of environmental protection, process water availability, and investment and cost. Their need is obvious when one considers the expected strong growth in the global demand for liquid fuels justified by the undoubted advantages of this form of energy for many uses.

9.3.4 Concern over the environmental impact

Hence the main constraints do not come from the resource side. In contrast, the consumption value of fossil fuels that could be reached by the middle of the century may result in significant gas emissions, particularly in terms of CO_2. Will mankind be able to cope with the threats it poses to its own environment? This is an issue far beyond the scope of this study, but that could well be one of the main limiting factors for the future, although no effective negotiation mechanisms exist today to tackle this inherently transnational issue.

Recently, energy has been found to be directly involved in major issues such as the accumulation of CO_2, the breakdown of the ozone layer, acid rain and urban pollution, as well as risks of nuclear accidents (Three Mile Island, Chernobyl, Fukushima), large-scale marine pollutions (Gulf of Mexico), extreme weather events (Katrina, Sandy).

International negotiations on the reduction of CO_2 emissions proceed at a pace well below those for CFCs, and we do not know yet if such drastic measures as were decided for the latter will be undertaken one day. In the meantime, in the absence of a scientific consensus, which seems impossible in the immediate future, and of economically viable alternative technical solutions, it is likely that international discussions will not succeed in the short term.

The risk of nuclear accidents remains a constant, even if the security in this area has always been a major concern for electricity producers. In the new Generation III or IV reactors that are currently being developed, emphasis is placed on this aspect.

To minimize the negative impacts of the energy sector on the environment, remember that three types of solutions can be considered:

- the rational use of energy (energy conservation);
- the use of "clean" technologies;
- the implementation of prevention and repair of damage.

9.3.5 Reducing the energy content of GDP

Since 1973, the demand management policies implemented by the governments of importing countries have significantly reduced the energy intensities in the world (1% per year since 1973). They continue to grow in the poorest countries, but we have seen that it is justified by their low level of per capita consumption.

Note that the average per capita primary energy consumption we have considered for the limit scenario (3.5 toe/capita) could correspond in the future to a consumption of useful energy much higher than today's, thanks to possible progress in energy conservation.

Efforts should focus both on the optimization of the supply system (refineries, and chiefly power generation), and on improving energy-using equipments.

The analysis of changes in the primary energy content of GDP needs to be done carefully because it includes very diverse factors, among which two in particular: the GDP structure and the effectiveness of the national energy system. Their evolution therefore reflects both the structural changes undergone by the national productive apparatus and technological progress in all sectors, especially in terms of energy conservation.

Extrapolation of the results observed in a given period is quite risky, the proportion attributable to each of these key factors being impossible to distinguish rigorously.

It is likely that the structure of GDP can only evolve within fairly small limits, once the main adjustments are done, so that it can be expected that the structural effect only applies once. In contrast, the diffusion time of energy-efficient technologies is all the longer as the fleet of equipment is important, so that the full effect of energy conservation policies will only be gradually visible, which suggests a sustainable continuation of trends. On the other hand, the energy conservation possibilities certainly remain considerable as many innovations are still possible although the easier to implement have already been implemented.

Finally, it is certain that prices play a role, at least in the long term, on consumer behavior, and so this exogenous variable will also have an impact on energy intensity.

Moreover, although the evolution, especially downward, of the primary energy content of GDP has been quite regular, since the second oil shock, the situation is a bit more complex in the case of the primary oil content, which in most importing countries declined significantly between 1980 and 1986, marking a clear break with a period of strong growth during the 1960s and until 1973, but stabilized again in a number of cases (or even increased, as in Japan), when prices were low again, then decreased from 2005, but more slowly than the energy intensity.

As energy intensities decreased faster, this meant that substitutions took place, with cheap oil replacing coal or natural gas. This probably stems from the fact that the important reduction of the primary oil content in the early 1970s was largely due to inter-energy substitution, which can more easily be reversed as many facilities have since been converted to at least two energy sources.

The return to oil then partially explains the rise in the global demand observed after 1985. We should therefore be very careful when trying to extrapolate past trends to produce forecasts in this area.

The graph in Figure 9.14 shows the evolution of the OECD energy intensity and primary oil content since 1970. For this group of countries, the two indicators have remained on a downward trend since the first oil shock, but this is not an immutable rule.

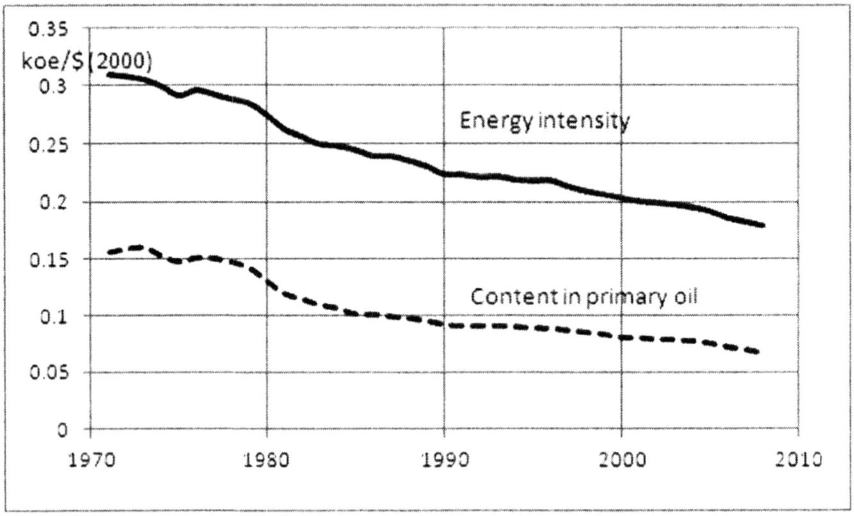

Figure 9.14 Evolution of OECD energy intensity and primary oil content.

9.3.6 A continued leading role for oil

We have seen that many degrees of freedom exist in an energy policy, and that each country seeks to maximize its gains, depending on its own resources and its economic and social situation. Oil being one of the most versatile and least costly (investment-wise) forms of energy, at least in terms of use, demand for oil is directly related to its price. When cost is high, consuming countries turn away from it and undergo austere energy policies, which bear fruit after a few years, but are challenged when prices fall.

Moreover, the amount of investments made by the world oil industry in exploration, production and maintenance of existing wells has an impact on prices because it determines the production capacity at a given time. But this level of investment is itself correlated with oil prices.

It should be noted that the structure of oil prices is characterized by a strong difference between production costs of conventional deposits and selling prices (2–3 dollars per barrel as against 60 to 80, or more). The high prices experienced by the end user do not reflect technical costs but various revenues, levied by both producer and consumer countries (taxes on petroleum products). It is therefore quite possible and inexpensive for producers to adopt dumping policies at regular intervals to regain market shares. Their profit margin being considerable, they are indeed easily led to set a price level where competition cannot really afford to economically produce substitutes (see section 8.4.3).

If such price changes are bearable by the major oil producers, they can however be devastating to the producers of unconventional fuels that are on the verge of profitability. The risk is great to see the development of these unconventional energy sources delayed with the exhaustion in priority of reserves at low cost, even if consumer countries continue to be in a situation of increasing dependence.

There is reason to be concerned about this situation as it can generate international instability and will most likely postpone the preparation of the needed long-term transition.

Even if it has recently slightly reduced, demand for oil has continued to grow steadily during last decades and oil trade remains much larger than that of other energy sources. Oil will therefore continue to play a leading role in the future.

A question often debated is whether the cause of the oil shocks must be attributed to the political action of OPEC or if it is due to a sort of economic determinism relative to the strong increase in demand for a finite resource (in the order of 8% per year before 1973, against 1% per year between 1983 and 2007).

Presumably, each of these theories holds some truth: the origins of oil shocks are neither purely political nor purely economic. Similarly, 1986's countershock and its amplified impact on OPEC can be partly explained by the economy, and partly by considerations of security of supply, and therefore political reasons.

In a prospective analysis of a field as complex as the world energy scene, simplistic arguments are not helpful. The future will be the result of combined economic and political factors.

9.3.7 Persistence of geopolitical constraints

It remains true that many problems of distribution arise and there is no guarantee that access to resources, whether financial or energy, will be free for all. The geopolitical nature of energy supplies has long been recognized as essential by most consumer countries, and the last forty years have shown that whenever the demand for oil is approaching the limits of supply capacity, the energy leaves the economic scene to become political. The continuous curve in Figure 9.15 shows, in millions of barrels per day, the worldwide excess oil production capacity between 1991 and 2008 (left scale), while the dotted curve displays oil prices in the U.S. (2005 WTI spot price, right scale) [ICPP, 2007]. We clearly see that periods of low excess capacity (about 1 Mb/d) are those when the oil price is higher. Let us recall that the world capacity was then close to 85 Mb/d.

While overall energy resources are not limited, those of oil are, and they are moreover geographically concentrated in a few regions, which further complicates the situation. Although its share in the global balance decreased (from 47% in 1973 to 37.5% in 1993 and 33% in 2011), oil demand remains strong, and resources limited ($R/P \approx 50$). Given its strategic importance, it will continue to weigh heavily on the energy scene, dependent on political crises such as the Gulf wars, or technical ones as capacity crises which occur at regular intervals.

Natural gas reserves are about the same size as those of petroleum, and are more geographically concentrated. However, the share of gas in the global mix as in international trade is smaller, which limits the geopolitical risk. If its role should increase significantly, the main question relates to its transport over long distances, as complex technical problems arise and as it is very vulnerable in times of crisis. However, the question of inter-regional gas market development is primarily linked to its indexation to the price of oil. Indeed, if this indexation persists, the risk of a shock is the same with gas as with oil, which may discourage consumer countries, taking into account additional constraints imposed by this highly volatile product. As mentioned in Section 3.8.5, the indexing of prices is however increasingly questioned.

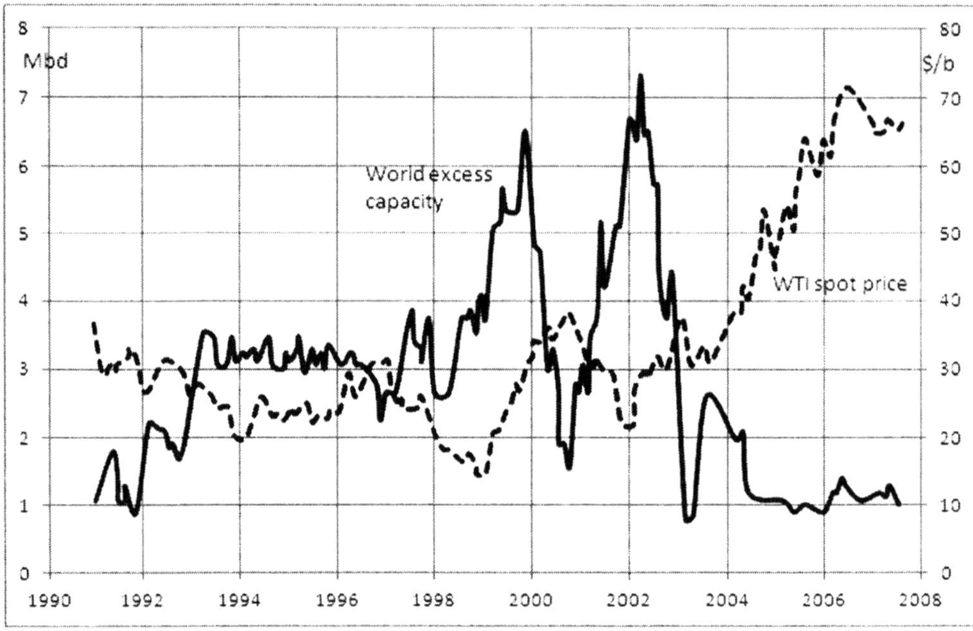

Figure 9.15 Excess capacity and oil prices.

Among geopolitical issues is the problem of the role of the Russian Federation as a producer (and consumer) of oil and natural gas, and uncertainties which exist about its share in world supply (and demand).

Another important issue is the evolution of the oil imports of the main consumers. We presented in Chapter 6 the dynamics of the major oil importers (U.S., Europe, rest of the OECD, India and China), and indicated that these two latter countries have continued to increase their imports despite higher oil prices since 2005, and that the United States imports very recently decreased, while those of the rest of the OECD, Europe included, have stabilized over the past ten years.

The overall outlook for the long term allows us to see in what context the shorter term changes occur. Two main questions come first: what role will oil play in the future, an issue that we discussed previous section, and will the observed decoupling between economic growth and growth in energy consumption be durable?

9.3.8 Continuation of a sustained technological development effort

In any case, a major technological development effort seems imperative. Whether for better use of existing oilfields (EOR), for discovering others, for developing non-conventional fuels, for reducing processing and distribution losses, for managing future renewable energy (solar, fusion, etc.), for reducing specific consumption (energy conservation), or for overcoming environmental constraints, technological innovation is seen as an essential pathway, which, in any event, will help make the transition easier and probably more socially acceptable.

Given that energy remains one of the necessary factors of production for economic development in all countries, technological innovation thus is one of the main keys to the future of energy.

In addition to its direct impact on the evolution of energy intensity, technological innovation has other advantages: firstly, it is a very interesting alternative to the conventional price adjustment, which may possibly eventually turn out to limit the adverse effects of oil shocks; moreover it is the only way to reduce the negative impact of energy consumption on the environment.

It should also be noted that a technological adjustment, as opposed to a price adjustment, generally has a positive impact on the economy as a whole, because it translates into job creation to produce new technologies, productivity gains and therefore industrial competitiveness, and into the development of products with high added value, etc.

Technological innovation seems to be the only way to develop energy systems with the characteristics of durability, reasonable cost, low environmental damage, adaptation to an increased demand for goods and services, etc., in other words technology that can help meet the long-term challenges while avoiding the short-term cyclical crises.

As we saw in the previous chapter, even a technologically mature sector such as energy is likely to experience major technical developments. The last decades have witnessed considerable progress made both in the extraction of oil and the development of new fuels, improved electricity production efficiencies, specific consumption reductions of many facilities, while related pollution was limited.

New technological changes are possible in the medium or long term, two of them being often mentioned:

- the development of hydrogen technology, a form of energy whose great advantage is not to generate any pollutant in the final combustion, but which poses many technological problems, particularly in terms of containment;
- controlled fusion, which would theoretically provide almost limitless quantities of energy, but that is not under control as of now.

Given the strategic nature of the energy sector, deeply nested with the entire world economy, states have quickly realized that they could not allow the market to act as the sole regulator. This is why it is complemented in most countries by a national energy policy, which usually pursues two main objectives: to increase the supply (through production investments and diversification of sources of supply) and reduce the demand (through greater energy efficiency and behavioral change). Technological innovation is an essential component of the energy policy.

To prepare for future changes, we must maintain a very open technological watch, avoiding to become trapped in monosectoral approaches: the emergence of a new energy system is likely, due to potential energy substitutions, to concern many other sectors that may appear well established. Therefore, it is necessary that technological progress be evaluated not only in terms of technical performance (efficiency, power, cost), but also increasingly taking into account the different external effects on which policymakers rely to carry out their decisions.

Let us emphasize one last point: political impetus is needed because the economic actors may tend to focus on the short term, but this impulse is particularly difficult to mobilize in times of crisis. As shown by the negotiations on the Kyoto Protocol, many countries in Annex 1 announced in Doha that they would not participate in the second phase, a result of the economic crisis.

REFERENCES

BP, BP Energy Outlook 2030, London, January 2012.

CME, Conseil Mondial de l'Énergie, Horizons énergétiques mondiaux 2000–2020, Technip, Paris, 1989.

Dessus B., Pharabod F., Jérémie et Noé, deux scenarios énergétiques à long terme, Revue de l'Énergie, n° 421, Paris, juin 1990.

Goldemberg J., Johansson Th. B., Reddy Amulya K.N., Williams R.H., Energy for a sustainable World, John Wiley and Sons, New Delhi, 1988.

IEA World Energy Outlook 2008, IEA, Paris, 2008.

van der Veer J., Grundy P., Shell energy scenarios to 2050, Shell International BV, 2008.

FURTHER READING

Spreng D., Flüeler Th., Goldblatt D.L., Minsch J., Tackling Long-Term Global Energy Problems: The Contribution of Social Science (Environment & Policy), Springer, 2012.

Annex 1

Pedagogical remarks

A1.1 INNOVATIVE PEDAGOGY

The Diapason modules are educational animated slide shows, each provided with a soundtrack. This is Information and Communication Technology (ICT) applied to education. Using this ICT application, one can present theories, methodologies or technologies to students.

These modules allow students to work by themselves at their own pace, alone or in groups, and access the oral explanations given by the teacher online at any time, in addition to written or electronic materials available to them.

They are structured in steps and sessions as well as in courses and trails. This makes the conception of rich educational environments easier, in particular in the form of alternate online sessions and classroom activities.

According to their own preferences, students may:

- either work primarily with the book, which can be easily read, or parts of the book by studying a list of sections suggested by the teacher;
- or start by taking the online course and use the book as a supplement and reference.

The content of the online sessions is a summary of that of the book, which is much more comprehensive. We therefore recommend that students start by viewing the Diapason sessions in order to get acquainted with the basic notions. Subsequently reading the book will provide them with additional information on the topics they are mainly interested in.

The combined use of Diapason sessions, other online resources (databases) and the book allows teachers to provide students with different learning activities suited to varied teaching contexts, mixing online work with collective activities in the classroom.

As explained in Chapter 1, students may work in small groups (typically four) in order to prepare an analysis of the energy situation in a country, with the aim of making an oral presentation to their peers and delivering a small written report, of typically ten or twenty pages. The ideal is of course when one of the students in the group is a citizen of the studied country.

The teacher provides students with data (national final energy consumption and energy production from IEA energy balances, and BP statistical review of world energy), while they must complement this information thanks to other sources, including from the studied country itself.

They are asked to analyze a number of indicators from the chosen country and group their changes in recent decades in tabular form: energy consumption, CO_2 emissions, per capita consumption, energy and carbon intensities, per capita GDP in current and constant values. The amount of time devoted to this activity should be about 12 to 18 h.

The case studies presented in Chapter 7 may be used as guidelines for their analysis. They can also be used as typical examples of analyses to be performed. For example, it may be interesting to ask students to explain the details of the anomalies in the evolution of some variables.

Experience has shown that this educational activity is met with great enthusiasm by students who would actually prefer to spend even more time on it. Having to perform the synthesis of a real national energy situation for oral presentation in front of their peers is extremely motivating and enables them to develop their capacity for synthesis.

Another kind of activity is to make a cross-comparison of several countries and to try and explain the differences in the policies implemented by them.

It is also possible to ask students to elaborate on the scenarios proposed in Chapter 9, refining them by introducing various groups of countries of similar socio-economic or energy characteristics, in order to set up new scenarios.

Obviously numerous other activities may be proposed to students depending on the available time and the objectives of the course. They may focus on some of the major issues at stake such as those we highlighted in Chapter 9:

- the levy on world wealth;
- the environnemental impact of energy;
- the reduction of the energy content of GDP;
- the future role of oil;
- the persistence of geopolitical constraints;
- energy for the rural populations;
- research and development efforts.

For such activities, teachers may ask groups of students to find additional information in the scientific literature and via the internet, in order prepare a small report on the topic, possibly together with an oral presentation.

A1.2 LEARNING OBJECTIVES

At the beginning of the course it is recommended to specify the educational objectives to the students, in particular by identifying the different types of knowledge they should acquire by the end of the course, such as the following.

A1.2.1 Memory

They should complete the course having perfectly memorized the following knowledge:

- definitions of key indicators (GDP, GNI, GDP, HDI, primary, secondary and final energy, energy intensity, CO_2 emissions and intensity);
- differences between energy reserves and resources;
- orders of magnitude of proved energy reserves.

A1.2.2 Understanding

They should have understood the following concepts:

- rigidity and implementation time of energy policies;
- international comparisons of key macroeconomic indicators;
- GDP structure;
- key development factors;
- problems specific to developing countries;
- environnemental impact of energy;
- regional distribution of energy consumption and production in the world;
- main features of international energy markets;
- impact of importing countries' energy policies on the oil market.

A1.2.3 Know-how

The following skills should have been mastered:

- calculate changes in energy intensities of different countries on the basis of available statistical data;
- analyze a national energy balance.

A1.2.4 Capacity for analysis and synthesis

Finally, students develop their skills of analysis and synthesis when they work on the project by applying all the knowledge acquired in order to analyze the energy situation of a country.

Annex 2

Energy units and conversion factors

Various units are commonly used to account for energy.

A2.1 BASIC UNITS

A2.1.1 Joule

The official unit of the International System is the Joule (1 J), equal to the work done by a force of 1 Newton whose point of application moves 1 m in the direction of the force.

However, this value is little used in practice.

A2.1.2 Calorie

Historically, the calorie has long been used to measure heat quantities. It represents the energy amount required to raise the temperature of 1 gram of water by 1 °C at the atmospheric pressure. As this value varies slightly depending on the initial temperature of the water, there are differences between the used definitions. We retain the following equivalence:

1 cal = 4.187 J (International Steam Table calorie)

A2.1.3 British Thermal Unit

An equivalent definition is used in the British system, still widely used especially in the United States: the British thermal unit or BTU represents the amount of heat required to raise by one degree Fahrenheit the temperature of 1 lb (pound) of water (0.4535924 kg).

With the same convention as for the calorie, we have the following equivalence:

1 BTU = 1055.6 J

This implies:

1 BTU = 251.984 cal

A2.1.4 kWh

The unit of power corresponding to the Joule energy unit is the Watt: a power of 1 W used for 1 second releases energy amounting to 1 Joule.

1 kilowatt-hour (kWh) is the energy amount released during one hour by a 1 kW power device.

Kilowatt and kilowatt-hours are units of power and energy most often used to measure electricity.

By definition:

1 Wh = 3,600 J
1 kWh = 3.6×10^6 J = 3.6 MJ

A2.2 DERIVED UNITS

To study the consumption and production of energy at the level of a country, Joule or calories are rarely used.

Due to the importance of oil in the energy balance of many countries, the amount of energy is frequently expressed as an oil equivalent: tonnes of oil equivalent (toe), kilograms of oil equivalent (koe), or barrels of oil equivalent (boe).

Other units are also used. This depends on the energy sources considered and their characteristics, which often vary a lot depending on the case. For natural gas, one uses cubic meters (m^3) or cubic feet (cf) under standard conditions of temperature and pressure (25 °C and 1 atm for m^3, 60 °F and 30 in Hg for cf). The equivalence between these two units is:

1 Nm^3 = 37.24 Ncf
1 Nm^3 = 38 MJ
1 cf = 1020 Btu

Primary or final energy is often reported in terms of millions or billions of tonnes of oil equivalent, or, in countries using the British system, in quadrillion Btu (quad):

1 quad = 10^{15} Btu = 1.055×10^{18} J

Conversions between these units are possible, but values may slightly differ depending on the authors. The IEA/OECD choose the following equivalences:

1 toe = 41.868 GJ = 39.68 MBtu = 11.63 MWh
1 boe = 5.8 MBtu = 0.146 toe
1 toe = 6.84 boe
1 quad = 25.2 Mtoe

Note that quite different toe/boe equivalence values may be found in the literature, including 1 toe = 7.33 boe, representing a difference of 7.14%, which can be explained by variations in the density and calorific value of the crudes used as references. For instance, BP uses a conversion factor of about 7 in the BP statistical review of world energy.

As tonnes are most frequently used, let us point out that one must distinguish them from tons, using the following conversions:

1 short ton (ton) = 2000 lb
1 metric ton (tonne) = 1000 kg
1 ton = 0.907185 tonne

A2.3 CONVERSION FACTORS

Equivalences below allow one to convert figures between the most frequently encountered units.

A2.3.1 Oil

- 1 tonne = 6.84 barrels
- 1 barrel/day for 1 year ≈ 50 tonne
- 1 toe = 41.868 GJ
- 1 toe = 11,628 kWh
- 1 bep = 5.8 MBtu

A2.3.2 Natural gas

- 1 m^3 = 38 MJ
- 10^9 m^3 ≈ 910 ktoe
- 1 cf = 1020 Btu
- 10^9 cf ≈ 25 toe

A2.3.3 Coal

- 1 t = 0.7 toe (coal)
- 1 t = 0.55 toe (sub-bituminous coal)
- 1 t = 0.25 toe (lignite)

A2.3.4 Electricity

- 1 MWh = 0.2606 toe (equivalent to production)
- 1 MWh = 0.086 toe (equivalent to consumption)

Subject index

For Product Safety Concerns and Information please contact our EU representative GPSR@taylorandfrancis.com Taylor & Francis Verlag GmbH, Kaufingerstraße 24, 80331 München, Germany

Batch number: 10726368

Printed by Printforce, the Netherlands